BEYOND STAR TREK

ALSO BY LAWRENCE M. KRAUSS

The Physics of Star Trek

Fear of Physics: A Guide for the Perplexed

The Fifth Essence: The Search for Dark Matter in the Universe

BEYOND STAR TREK

PHYSICS from Alien Invasions to the End of Time

LAWRENCE M. KRAUSS

BasicBooks
A Division of HarperCollins*Publishers*

 For information address HarperCollins Publishers, Inc., 10 East 53rd Street, New York, NY 10022.

FIRST EDITION

Designed by Elliott Beard

Library of Congress Cataloging-in-Publication Data

Krauss, Lawrence Maxwell.

Beyond Star Trek : Physics from alien invasions to the end of time / by Lawrence M. Krauss. — 1st ed.

p. cm.

Includes index.

ISBN 0-465-00637-X

1. Space sciences. 2. Life on other planets. 3. Space flight.

I. Title

QB500.K64 1997

001.9'01'53—dc21 97-31127

---

97 98 99 00 01 ❖/RRD 10 9 8 7 6 5 4 3 2 1

In Memory of Carl Sagan
1934–1996

"Make it so!"

—Jean-Luc Picard

CONTENTS

PROLOGUE

These are the days of miracle and wonder.

—Paul Simon

I have been asked innumerable times since the publication of my last book, *The Physics of Star Trek,* to talk about the relationship of science to science fiction. I think the connection is a simple one: We are all inspired by the same questions.

I also believe that the questions that scientists and writers of science fiction wonder about are essentially universal and time invariant. They are the subject of every age's fascination, reflected in its literature, art, and drama, and its science. The specific miracles change with time, as we learn about the world; as certain mysteries are unveiled, others are born. Think about a vibrant flower. Could such a wonderful thing really have evolved from primordial sludge? Yes. But let's go beyond this rather tired question and examine the flower further. It may have a beautiful pattern visible only in ultraviolet light, which a bee can sense. Who ordered that? Or think about the myriad chemical reactions going on in the bee's eye, which turn individual packets of pure energy into the same visual picture each time the bee scans the flower, in spite of the fact that these reactions are governed by probabilistic laws and the very molecules that respond to the light cannot be said even to exist in any specific state before, and sometimes after, absorbing the light. Deep inside the bee's brain and our own, the mysterious quantum-mechanical universe turns into the classical, predictable universe. How? And why are *we* self-aware and not the bee? Do we represent the only full consciousness in the universe? Are there extraterrestrial intelligences conscious of us now? How will we ever know?

All the miracles of our own existence and others' can be expressed in scientific terms. But the issues are just as engaging to anyone who simply wonders, "What if . . . ?" However, while the best science fiction arouses our interest by capturing the drama and excitement inherent in the "What if . . . ?" questions, it generally leaves the answers hanging. Modern science holds the key to knowing what is possible and what isn't.

Celebrating the connection between science and popular culture is therefore a natural way to set out the ideas that drive the modern scientific enterprise. Moreover, it can be a lot of fun. I have chosen here to go beyond *Star Trek*—to range over a larger collection of examples and anecdotes, and to treat issues that more widely permeate our culture. I'm not abandoning Trekkers, just, I hope, opening the door for an audience who may not stay up to watch the reruns every night. I hope, too, that those readers who may have been waiting for *The Wrath of Krauss* will not be disappointed. The inspiration for much of what I will discuss here has been derived from matters raised in thousands of e-mails and letters, and in conversations I have had with readers over the past 2 years—and, as you will see, *Star Trek* is never far away. The enthusiastic response to the previous book has been a great gift for me. I hope this one will be an adequate, if partial, repayment.

So, buckle up. Here we go again.

BEYOND STAR TREK

SECTION ONE

They'll Be Comin' Round the Mountain . . .

SCULLY: There's a marsh over there. The lights . . . may have been swamp gas. . . . It's a natural phenomenon, in which phosphine and methane rising from decaying organic matter ignite, creating globes of blue flame.

MULDER: That happens to me when I eat Dodger Dogs.

CHAPTER

ONE

Choose Your Poison

It's just that in most of my work, the laws of physics rarely seem to apply!

—*Fox Mulder*

A dark, ominous shadow descends over your house. The furniture starts to rattle, the walls and ceiling vibrate, and you hear a strange whistling in your ears. You rush to the window to see what's causing all the commotion. Only 5,000 feet off the ground, a huge black disk at least 15 miles across floats motionless in the sky, blotting out the sun, darkening the entire neighborhood. You run to the kitchen sink and splash cold water on your face. Surely this can't be happening! Back to the window once more, and the massive object is still there. You scurry out to the garage to get away, then you remember something. Hurrying back to the house, you pick up the phone to call your daughter's school, but the line is dead. You lose bladder control. The realization terrifies you. Aliens have arrived! As you begin to black out, your last thought is, *I am about to become toast!*

■ ■ ■

Hold on! While F14s or computer viruses or even H. G. Wells's microbes might not be able to protect us from the sheer terror generated by the attack of a 15-mile-wide floating saucer, Isaac Newton would—sort of. Newton's laws would ensure that you'd probably be dead before you had time to get terrified. Even 350 years after the fact, Hollywood still has to get past Newton before it can indulge in all the fancy stuff. Alas, the aliens piloting the Mother Ship in the blockbuster *Independence Day* seem to have skipped that semester back home. . . .

What instead might actually transpire if we were visited by the Mother Ship and her children reads more like a scenario for the Salem witch trials.

DEATH BY DROWNING

A Mother Ship full of aliens bent on ending life on Earth may not need to send out a squadron of huge flying saucers in order to destroy our major cities. Long before the first shadow fell on the Empire State Building or the Hollywood sign, New York might be underwater and Los Angeles could be leveled by earthquakes. Early in *Independence Day*, the telemetry tracking the approach of the Mother Ship reveals that it is almost ¼ the mass of the Moon. Before it releases its squad of death saucers, the mammoth ship pulls into a geostationary orbit above the Earth—the same sort of orbit the *U.S.S. Enterprise* uses to visit a new planet. In such an orbit, a spacecraft or a satellite moves at the same rate as the planet rotates, so that it always stays directly above the same spot on the planetary surface. The large communication satellites that transmit our international messages, as well as the network of Global Positioning navigational satellites that guide our airplanes and well-equipped trekkers (the terrestrial wilderness type), sit in such orbits.

Newton's law of gravity determines how high such an orbit must be, regardless of the object's mass. It is one of the many miracles of the law of gravity that any object, no matter how heavy, must orbit at exactly the same speed as any other object at the same distance from Earth. (If that weren't the case, NASA would have to design a different trajectory for every space shuttle, depending upon the weight of the astronauts inside.) The distance from Earth for an object in geostationary orbit is about 22,500 miles, or almost 1/10 the distance from Earth to the Moon. At 22,500 miles up, the gravitational attraction on the Earth of an object the mass of the Moon would be 100 times stronger than the Moon's gravitational pull; since the Mother Ship is 1/4 the mass of the Moon, its gravitational pull on the Earth would be 25 times that of the Moon!

What would this do? Well, one effect might well be to close down Wall Street, because much of New York City would probably be awash. The tidal forces provoked by an object as massive as the Mother Ship would cause a catastrophic rise in sea level in various places on the Earth. At the same time, the unaccustomed tidal stresses on the Earth's crust would undoubtedly induce earthquakes and volcanic eruptions in sensitive areas around the globe. Moreover, the very motion of the Earth through space would be affected, producing unpredictable effects, including possible climatic variation. When an object as heavy as 1/4 the mass of the Moon is in close orbit above the Earth, it causes the Earth to move back and forth in response—once again, because of gravity. Adding a third massive body, with its additional gravity, to the Earth-Moon system would change the system's dynamics in possibly chaotic ways.

Indeed, if the evil aliens were particularly patient—and why shouldn't they be?—they might choose to orbit the Earth in the direction opposite to its present direction of rotation. The tidal pull of the Mother Ship would then slowly serve to brake the Earth's rotation rate, lengthening the day or getting rid of it all together! In just such a way, the length of the Earth's day has been slowing due to the Moon's pull. Eventually (on a cosmic timescale), the Earth's rotation period would precisely match the orbital period of the Moon, so that one Earth day would be almost a month long. Imagine how hungry you would get between lunch and dinner.

Whether or not its crew chooses the slow route or the fast one, the Mother Ship could wreak devastation on Earth by astute choice of orbit, without doing anything more than being there—much easier than risking battle with terrestrial aircraft and missiles.

. . . OR BONE-CRUSHING

So much for the Mother Ship. The mammoth 15-mile-across flying saucers, whose shadows over the White House, New York, and Los Angeles produced some of the most memorable movie images of 1996, would also pack quite a wallop without firing a single shot.

Let's first imagine how much a ship 15 miles in diameter—and, say, 2 miles in height—would weigh. Now, the ship is not solid, of course—there has to be interior space for the aliens to move around in. So let's assume that 1/10 the volume of this object consists of structural elements and the aliens themselves, and that the rest is essentially just air (or some comparable gas); and let's give them the benefit of the doubt and assume that the solid material is lighter than steel—say, with the density of water (1 gram per cubic centimeter). I estimate that such an object would weigh approximately 100 billion tons.

That's pretty heavy. But an airplane is pretty heavy, too, and it flies. Well, there's a big difference. We can figure out how big by asking what kind of upward force would be required to hold this gigantic craft against the downward pull of gravity. Note that we can ask this question independent of whatever exotic physical mechanism the ship uses to levitate, be it as "conventional" as fusion-powered thrusters or as far out as antigravity. We express the question in terms of the pressure the ship would need to exert on the atmosphere below it to keep it aloft, given its weight. Dividing the weight of the craft by the area of its disk, one gets a pressure of about 450 pounds per square inch directly below the craft—or about 30 times the normal atmospheric pressure we feel at sea level.

We tend to ignore the pressure of the atmosphere; after all, we are surrounded by it all the time. But the Earth's atmospheric pressure is really remarkable, when you think about it. At sea level, the atmosphere exerts a pressure of 15 pounds on each and every square inch of your body. That's about 150 pounds on the palm of one hand! Why don't we feel it? Because our bodies are in what is called hydrostatic equilibrium with the atmosphere—that is, the fluids and gases inside our bodies exert a pressure outward equal to the pressure inward from the atmosphere. Change the balance, however, and dramatic effects ensue.

The effects of atmospheric pressure were demonstrated as early as 1657, by Otto von Guericke, the mayor of Magdeburg, who invented the vacuum pump. He fitted two copper hemispheres the size of backyard barbecue kettles together to form a sphere. The two hemispheres weren't soldered or glued together and could easily be separated. But when he evacuated the air in the sphere, so that the atmospheric pressure outside the sphere was not balanced by the air pressure from inside, two teams of eight horses apiece were unable to pull the hemispheres apart! Fifteen pounds per square inch adds up.

Recall that the pressure exerted downward by one of those flying saucers would be about 450 pounds per square inch. That means an extra weight of about 30 tons per square foot on every object on the surface just beneath it. A normal building will collapse from an overpressure of about 5 atmospheres, or some 5 tons per square foot, which is the overpressure produced by an average nuclear weapon at a distance of about 10 kilometers. Forget about giant weapons belching fire: to flatten major cities, the huge disks could just sit there in the sky! Of course, this wouldn't have made for spectacular previews of coming attractions.

Why, you may ask, don't conventional aircraft crush people and buildings as they travel above them? Well, aircraft are not really very heavy compared with the weight of the atmosphere. A 100-ton aircraft measuring 100 feet long by 10 feet wide needs to exert a downward force of less than a pound per square inch on the air below it to stay aloft. More important still is the fact that a plane's cruising altitude is high in relation to its size. As the airplane rises, the pressure it exerts on the atmosphere below spreads over a larger and larger region, so that it is significantly attenuated by the time it's transmitted to the ground. When you are far below the craft, you are unlikely to feel anything at all (except the noise of its engines). The same would be true for the giant alien spacecraft, if they were so far above the ground that their altitude was much larger than their breadth—but then they would appear as inconsequential disks in the sky, not the towering behemoths of *Independence Day*.

. . . OR TRIAL BY FIRE

Let's say we're lucky enough to survive the floods and the quakes and the crushing pressures, and we then send up a huge force of F14s led by a young ex-fighter-pilot president and actually manage to disable the saucers. Suddenly we're not so lucky anymore!

How much energy is released when a single spacecraft of this size plummets to Earth from a height of, say, 1 mile? I conservatively estimate it to be something like 10,000 times the energy released by the nuclear weapon that destroyed Hiroshima. I'm not sure that in this case the winners would feel much like celebrating. Remember that the impact on Earth of a single comet or asteroid—thought to be no larger than such a spacecraft, albeit traveling at a faster speed—was probably responsible for wiping out much of life on Earth at the end of the Cretaceous period, 65 million years ago. And remember that a lot of the huge saucers were downed in the *Independence Day* victory.

In fact, the power required simply to move such a spacecraft into our atmosphere would be devastating. Considerations of energetics allow one to calculate that to accelerate a craft of this size in a minute to a speed of 3 miles per second (say, about half the escape velocity from Earth) would require a power expenditure during that minute of something like 50 billion billion watts—about 300 times more than the power received on Earth from the Sun and a million times the average power used by all of humanity in our daily existence. The heat radiated by many such spacecraft would be enough to make it feel more like Doomsday than Independence Day.

Which brings us back once more to the good old Mother Ship. How much energy would be needed to slow down or speed up an object ¼ the mass of the Moon so that it could enter or leave Earth orbit? The amount is almost unfathomable. I have tried hard to think of something that would adequately represent what would be required, and I hope this works: If it took the Mother Ship's engines an hour to slow the craft down, the energy radiated by these engines would be almost 10 times the entire luminosity of the Sun during this period! Imagine a Sun shining on us not from 93 million miles away but from a mere 22,500 miles away. The intensity of the radiation would be about 25 million times stronger.

Toast? You better believe it!

CHAPTER TWO

To Be or Not to Be

The infinite quietness frightens me.

—*Blaise Pascal*

Our first contact with aliens need not be quite this menacing. One of the reasons I have enjoyed watching *Star Trek* in its various manifestations is that it presents a hopeful view of the future. Zefram Cochrane's fanciful first romp in warp drive, chronicled in *Star Trek VIII: First Contact,* was followed almost immediately by a benign encounter with Vulcans and an invitation to join the Federation. Given that the resources required to make the kind of interstellar voyage chronicled in *Independence Day* are so much greater than whatever one might immediately gain by plundering our planet, I doubt that anyone making the trip would initially be bent on conquering us. That might come much later . . . after they got to know us.

Aliens are cropping up all over, witness the recent successful release of *Independence Day* and *First Contact,* and the rerelease of George Lucas's *Star Wars* trilogy; by the time this book appears, the lineup of alien-visitation films will also have included at least four big-budget epics, including one based on *Contact*, a best-selling novel by the late astronomer and science popularizer Carl Sagan. In a distressing demonstration that life imitates art, religious cults based on the existence of aliens have sprouted up. Comet Hale-Bopp has claimed the lives of thirty-nine believers, who saw salvation in the supposed advent of an alien spacecraft. Earlier, science fiction writer turned religious prophet L. Ron Hubbard built a large religious empire based on his notions of long-dead extraterrestrial civilizations. It seems you can take out an insurance policy against abduction by aliens, and at least 4,000 people have so far put their money down, although no abandoned loved ones have yet collected. But I must admit to seeing little difference between the fanciful myths of true believers of the Heaven's Gate variety and those of more orthodox fundamentalists. (For example, it seems just as likely to me that there was a spacecraft hidden behind Hale-Bopp as that an ancient patriarch named Noah sheltered all known species of animals from a globe-girdling Flood in a giant ark.) The solace that people appear to obtain from the idea that we are not alone in the universe is powerful. As far as I can tell, Fox Mulder's *X-Files* mantra "I want to believe!" is widely shared.

It's easy to see why. Solitude in the vast expanse of empty space is unnerving, as the seventeenth-century French mathematician and philosopher Blaise Pascal put it (see the epigraph to this chapter). Infinite quietness is indeed frightening. The human yearning to fill the cosmic darkness with a Divine Presence—or at the very least with kindred forms of life—is as natural as the search for warmth and light in the primeval woods. What could be more exciting, and more comforting, than the discovery of some far-distant cousins in the universe? In the course of writing this book, I asked a number of eminent physicists and cosmologists to give me one question about the universe to which they would like a definitive answer. "Is there intelligent life out there?" was the response of Nobel laureate Sheldon Glashow.

We have all thought about that. In whatever way it happens, first contact would alter our civilization more dramatically than any other single event in human history. At a recent conference in Naples on the possibility of extraterrestrial intelligence, I gave a talk centered on the physics of the subject. At the same conference, George Coyne, who heads the Vatican Observatory, spoke about the challenge to Christian theology presented by the possible existence of extraterrestrial civilizations. His talk reminded me of the remark of the unarguably devout twelfth-century Jewish philosopher Maimonides, in *The Guide of the Perplexed,* that while the Scriptures were true, if the results of science disagreed with one's scriptural interpretation then one would have to reexamine that interpretation. After Coyne's talk, I asked him what seemed the inevitable question: Might theologians who address this issue have to conclude that the existence of extraterrestrial intelligent life is incompatible with the tenets of the Catholic faith? He answered that they very well might. My own feeling is that the discovery of extraterrestrial life would be far more jolting—and not just to orthodox Christians—than was the revelation that the Earth is not the center of the solar system. It is my long-held conviction (and clearly that of the producers of such films as *Contact*) that the discovery of aliens would surpass the Copernican Revolution in its consequences for our understanding of our own existence and for the persistence of our present belief systems.

So, how would you know a successful interstellar visiting spacecraft if you saw one? How would it behave? Thinking about such questions is a useful precursor to determining how we ourselves might one day carry out the mission of the *U.S.S. Enterprise* "to explore strange new worlds, to seek out new life and new civilizations, to boldly go where no man has gone before." The problem of how to recognize a vehicle from another world turns out to be a little more subtle than one might guess.

The traditional notion has been that UFOs don't behave like rockets or planes (this is, after all, what makes them UFOs). Strange lights that flit unlikely distances back and forth across the sky, like the dazzling display in Steven Spielberg's *Close Encounters of the Third Kind,* are typical. More recently, in one of the early episodes of *The X-Files*, the ardent UFO hunter and FBI agent Fox Mulder finally gets to see some real UFOs in a secret air force installation somewhere in the Southwest (could it be Area 51?), and these vehicles do just what UFOs are supposed to do—namely, everything our own aircraft can't. Mulder and his colleague Dana Scully are astounded by a series of bright disks moving at incredible speeds through the skies above the remote base, turning at 90-degree angles on a dime. Like many of the action sequences in *The X-Files*, this one evokes episodes in the canonical UFO literature. Well, *Catch-22* comes to mind, too. While a UFO might be defined as something that moves though the air in a manner unlike that of a conventional rocket or plane, I would argue that this is precisely how a genuine UFO would *not* behave!

Let me offer the following asides: One of the oddly appealing things about *The X-Files* is that it makes no concessions to reality. And as in all successful dramas, you identify with the characters; that identification is really what compels you to watch. Fox Mulder is the earnest New Age searcher, trained as a psychologist, always willing to be skeptical of the laws of physics and much less willing to be skeptical of his long-held beliefs. Dana Scully, the more rational "skeptic" of the two, was trained as a physicist—no less!—before her stint in medical school, and her gender constitutes a wonderful reversal, as far as the usual run of TV is concerned. I will be forever grateful to the series' producers for giving us this role model of an intelligent, attractive, and relentlessly pragmatic female physicist. She is the foil to Mulder's ineffable eagerness. She is always there to ask, Why? And she sometimes does.

In the UFO episode just described, it turns out that what Scully and Mulder have seen are alien spacecraft piloted by crack air force test pilots. The pilots can't handle the strain of racketing around in these unfamiliar ships, and they start disappearing from sight. Well, it's indeed likely that terrestrial pilots wouldn't be able to hack it; however, neither would the alien spacecraft—and Scully, a physicist, probably should have known as much.

Let's go back to Newton and briefly consider the stresses induced when your average UFO—traveling at, say, twice the speed of sound—makes a 90-degree turn. The speed of sound in air is about 750 miles per hour, or about 350 meters per second, so imagine that we are observing an object traveling at 700 meters per second and we see it turn 90 degrees. In other words, it suddenly stops traveling forward and now moves sidewise at a right angle; in effect, it comes to a halt and then resumes travel in another direction. What kind of force would be required to make a craft moving this fast stop on a dime? To be generous, let's say it takes 1/10 of a second for the vehicle to stop and change direction—a short enough time so that you might perceive it as instantaneous. Well, the deceleration of the spacecraft performing this maneuver would be about 700 times the acceleration that gravity produces in a falling object at the Earth's surface. In the language of the G-forces, familiar to aircraft pilots, aficionados of space exploration, and readers of my previous book, this means that the occupants will feel a force of 700 Gs. I remind you that the maximum G-force people can experience and survive for short periods is only 8 Gs or so. Experiencing 700 Gs would be the same as having a 70,000-pound, or 35-ton, weight pressing down on your shoulders (more or less the same force you would feel from the increase in atmospheric pressure due to the visiting saucers in *Independence Day*).

What effect would such a force have on the craft itself? Well, imagine a plane suddenly losing engine power at, say, 1,000 feet and falling to the ground. If it makes a crater a meter deep, I estimate that the G-force experienced by the plane during impact is about 2,800 Gs. Judging from what most plane crashes look like, I would suggest that no craft made out of mere metal would be likely to survive the *X-Files*-type aerobatics for long.

But you may argue that UFOs are not made of mere metal. The advanced civilizations that create them have made them out of superstrong materials. Well, OK—but what about the aliens themselves? Would they be able to withstand those levels of G-force? I don't see how, unless they evolved in an environment that produces 40-ton raindrops.

Be that as it may, what is the point of designing a spacecraft to perform right-angle turns and other such exotic aeronautics? As we will discuss, a voyage from another world is a demanding one, and at least 99.999 percent of the time will be spent in space. It's unlikely that any alien craft will be tailor-made to behave as an acrobatic sports plane in the Earth's atmosphere. Remember the *Apollo* missions to the Moon? (If you are over forty you ought to, and if you are under forty you may well have seen the estimable *Apollo 13*.) The mission's lander—the LEM, or Lunar Excursion Module—was spectacularly unaerodynamic. Why? Because its chief job was to descend from the orbiting command module to the lunar surface, where aerodynamics is irrelevant because there is no air. Our present space shuttle is designed more like an airplane, but that's because it has to spend a substantial and important part of its time reentering the atmosphere.

We tend to anthropomorphize aliens, and this may well have led us to "humanize" their spacecraft as well. For most of the past century, we've been used to traveling in the air, so it seems natural to imagine that craft from other planets must also be designed for air travel. Planes bank when they turn because they have to: they operate by using air pressure—that is, they fly because the air pressure above the wings is less than that below the wings. So to turn right, they have to bank, by raising their left wing and lowering their right one, which tips them rightward. In space, where wings do not figure in propulsion, the main reason to bank is removed. Yet the *Enterprise* and Han Solo's *Millennium Falcon* nevertheless always bank. Why? Well, the answer is the same as that for another question I'm sometimes asked: "Why does the *Voyager* lift its warp nacelles just before going into warp drive?" Simple: It looks good.

Since the summer of 1947—the same summer as the famous sighting at Roswell, New Mexico—when Kenneth Arnold, a commercial pilot, thought he saw a formation of silvery disks above Mt. Rainier and subsequent newspaper stories dubbed his visions "flying saucers," saucer-shaped vehicles have been the ship of choice for witnesses of alien visitation. Why not? After all, a spinning disk is satisfyingly stable—it can generate lift and it resists tipping over. Moreover, as an astute editor once remarked to me, "Isn't it uncanny that flying saucers were observed before Frisbees were invented? We now know that Frisbees are great at moving through the air. How could the early UFO observers have guessed this fact?"

Spinning disks are stable indeed, and Frisbees fly well. But both these facts are largely irrelevant where spacecraft are concerned. In the first place, we all know what happens if you are inside an object that's spinning at any significant rate. You're thrown against the outer wall. (You also tend to get sick, especially if you look out the window at scenery that isn't spinning.) While this is precisely the mechanism we will one day use to produce artificial gravity on spaceflights of long duration—as Arthur C. Clarke and Stanley Kubrick wonderfully depicted in the classic *2001*—a small craft spinning as rapidly as the saucers on TV would likely immobilize its crew against its perimeter. And simply causing the outside of the hull to spin won't do, either, since in order to be stabilized by rotation, most of a vehicle's mass has to be spinning.

Finally, as I've noted, interstellar (and even our own interplanetary) spacecraft are designed for traveling in space. A Frisbee flys well because of its aerodynamic properties. The spinning not only gives it stability but makes the air pressure less above the Frisbee's surface than below it. Where there's no air, this effect is useless. In the near vacuum of space, a Frisbee or any other saucer shape would perform as well as a flying pretzel. Should we expect an invasion by flying pretzels? Well, the best answer comes from trying to imagine what we ourselves would build. Whether visitors from space want to conquer us or invite us to join their federations, they will need to have solved the same problems we face if we are ever to escape our ties to the Earth.

CHAPTER

THREE

To Boldly Go . . . If We Can Afford It

Space is big. Really big. You just won't believe how vastly, hugely, mind-bogglingly big it is. I mean, you may think it is a long way down the road to the chemist, but that's just peanuts to space.

—*Douglas Adams,*
The Hitchhiker's Guide to the Galaxy

I was recently in the city of Geneva, and the words of a famous former resident, Jean-Jacques Rousseau, came to mind: "Man is born free, and he is everywhere in chains." More than 20 years have passed since humans last set foot on any object other than the Earth, and not even another manned voyage to the Moon is in our near future. Mars beckons us with tantalizing new hints of extraterrestrial life suggested by recent analyses of Martian meteorites, and recent images from NASA's *Galileo* spacecraft suggest that buried beneath the frozen surface of the Jovian moon

Europa is organic slush and perhaps even an ocean—a primordial breeding ground for life. Yet the possibility of human voyages to the Red Planet or the moons of Jupiter anytime soon seems remote. It is the threshold of the twenty-first century, and as a species we remain as Earthbound as ever. To those who yearn to break free from our terrestrial chains, our circumstances border on the tragic.

Our imprisonment is in stark contrast to the pictures bombarding us on the big and little screens, where beings voyage among the stars with impunity, using fusion, warp drive, hyperdrive, wormholes, antigravity, and whatever else pops into the fertile minds of the scriptwriters. So, what's our problem? How can we get from here to there? Well, at the heart of it all—even beyond the issue of what is physically plausible and what isn't—lies the matter of money. As pedestrian as it may seem, the chief factor limiting our ability to get a manned spacecraft even only to Mars and back again, much less to Alpha Centauri (the nearest star system, only some 4 light-years away), is that we cannot finance a mission involving a ship big enough to accommodate the needed fuel and a reasonable number of astronauts on a voyage of long duration.

In real life, and sometimes in science fiction, money determines the difference between what may happen, even in principle, and what does happen. I remind you that it was money, or rather the lack of it, that led Gene Roddenberry to invent the transporter allowing the *Enterprise* crew to "beam down" to planets, since he didn't have the budget to depict the landing of a spacecraft in the course of each episode. Finally, after 30 years, confident of a seventh surefire hit movie and a third spin-off hit television series, Paramount showed us the *Enterprise* crash-landing on a planet, and Kathryn Janeway's *Voyager* has also landed, a bit more smoothly, in a number of episodes. Nothing turns on a screenwriter's imagination like money—witness the words attributed to Kevin Smith, writer of the new *Superman* movie, due out in 1998:

"The budget is big. God almighty, it is big!" As far as real spaceflight is concerned, money translates not into production values (at least for those of us who believe that NASA really did put men on the Moon and didn't stage the whole thing on a Hollywood back lot) but into *energy*. Energy, in turn, means fuel.

This aspect of our problem may seem baffling at first. After all, two decades ago we were able to rocket a manned command module, complete with LEM, to the Moon and back; surely, rocket engines have not become less powerful since that time! Of course, Mars is about 1,000 times farther from the Earth than the Moon is, which appears to suggest that going there at the same speed would take 1,000 times longer, or almost 10 years one way—too long for any manned mission given the present state of our space technology. But the Earth is barreling around the Sun at something like 20 times the speed at which *Apollo* journeyed to the Moon. Thus, a Mars-bound spacecraft leaving Earth orbit will be launched from a platform already moving at considerable velocity relative to Mars. If one uses Earth's solar-system velocity as a springboard to propel a rocket to Mars, a one-way trip would take no more than six months to a year, assuming the craft traveled away from the Earth at only 2 or 3 times the speed of *Apollo*.

So, I repeat, what's the problem? Well, while the aforementioned increase in speed may not sound like much, it comes at a high cost. To understand this, we have to remember how a rocket works. Rocket propulsion depends on the law of physics called Conservation of Momentum. Put simply, this law states that if I throw something away from me, I will recoil in the opposite direction. Rockets "recoil" forward because they throw mass out their back ends. The speed with which a rocket is propelled forward depends on three factors: the speed at which the propellant leaves the back end, the mass of the expelled propellant, and the mass of the rocket and the fuel remaining on board. Thus, for example, an inflated balloon that is not tied off will fly forward if

I let it go, because it expels air quickly out the back. If the balloon were not so light—say, if it were made of concrete—it wouldn't go anywhere. Similarly, if the balloon is not inflated very much, so that the walls of the balloon are hardly stretched and the air is expelled out the back very slowly, it won't go anywhere.

Where balloons are concerned, one doesn't worry about the additional mass represented by the air inside. Not so for rockets: they require so much fuel that its additional weight cannot be ignored. And there's the rub: If I want my rocket to move faster, I have to throw more propellant out the back; but if I have to throw more propellant out the back, I must start out with more fuel aboard the ship. But if I start out with more fuel aboard the ship, I must expel a little more propellant than I would have to get the ship (plus fuel) moving in the first place. But that means I have to bring along more propellant . . . and so on and so forth.

The Greek philosopher Zeno faced a similar problem two millennia ago when he tried to add up an infinite series of numbers. The resolution is still the same: as long as the increment I keep adding gets smaller fast enough, even an infinite series of terms can have a finite sum. In this case, do the fuel increments needed get small enough fast enough? It turns out that the answer is yes—at least as long as one is traveling well below the speed of light; when you approach light speed, the effects of relativity begin to complicate things. Nevertheless, the final total amount of fuel required depends sensitively—in fact, exponentially—on the final speed of the ship relative to the speed at which propellant shoots out the back of the ship.

As this final, cruising velocity begins to exceed the speed of the propellant, things get unwieldy. Increasing the final velocity of a rocket from 1 to 2 times the speed of the propellant out the back requires 4 times as much fuel. But increasing the final speed to 4 times that with which the propellant leaves the ship will increase the required amount of fuel by a factor of more than 30!

In this case, the initial mass of the ship plus fuel would be about 55 times the mass of the ship without fuel.

In practice, things are even worse than this "rocket equation" predicts, since a ship designed to carry a disproportionately large volume of fuel will doubtless have to be sturdier than it would otherwise have to be, and will therefore weigh more. It is generally impossible to carry enough fuel to move a ship faster than 3 or 4 times the velocity of the propellant.

It gets worse still. A round-trip voyage even to the nearest planets would be, at best, a multiyear proposition. You must therefore design a spacecraft that can adequately house, feed, and provide a breathable atmosphere for astronauts for an extended time period. Such a spacecraft would have to weigh substantially more than an *Apollo* capsule. Since the total amount of fuel required is a fixed multiple of the spacecraft weight, this means that the net fuel requirement for a mission to Mars would be many times that associated with a voyage to the Moon, even if faster speeds were not needed.

Then there's the matter of getting back. Mars has a stronger gravitational field than does the Moon, so to achieve a trajectory back to Earth you would have to carry a comparable amount of propellant for the return trip, in order to build up a velocity, relative to Mars, comparable to the velocity one had to build up relative to the Earth to get to Mars in the first place. This means that the ratio of fuel required for the return journey, relative to the mass of the now lighter spacecraft, is similar to that required for the outbound journey.

However, if this fuel is brought along on the outbound journey, then it must be added to the initial mass of the spacecraft before you can calculate the initial fuel requirements. To get an idea of the problem, imagine that the fuel required to achieve a sufficient velocity is 5 times the mass of the spacecraft with an empty tank. If you need a comparable ratio to get up to speed on the return journey, then you would need to land on Mars with a spacecraft that weighed 6 times the mass of an empty spacecraft—that is, the mass of the empty spacecraft plus 5 times the mass of the spacecraft in fuel required for the return journey.

This would mean that the mass of the spacecraft plus fuel at takeoff from Earth would have to be 36 times the mass of the empty spacecraft! This accounts for the mass of spacecraft and the fuel for the return voyage, plus 5 times this total mass for the fuel required for the outbound voyage. I want to emphasize this point, since a somewhat similar estimate in my last book—associated with the mass of fuel required for the *Enterprise* to get to half light speed and then come to a stop—generated more mail than any other single item. The total amount of fuel needed is not 6 *plus* 6, or 12, times the mass of the empty spacecraft; it is 6 *times* 6, or 36. Pretty soon, to paraphrase former Lockheed-Martin rocket engineer Robert Zubrin, you end up with *Battlestar Galactica*! In light of this, perhaps the monstrous flying saucers in *Independence Day* were not all that unrealistic; they might have needed to be that big just to carry the necessary fuel!

The preceding scenario more or less governed what happened when NASA first officially considered a manned mission to Mars, in 1989. The price tag for the Monster Ship? Between $400 billion and $450 billion! At this cost, a human mission to Mars in our lifetime would remain a pipe dream—and this project pales in comparison to the requirements of getting beyond our solar system, out to other stars. Every aspect of such a mission would exacerbate the fuel problems I have mentioned.

The point is that our galaxy is *really* a big place. The distance even to the closest stars is many thousands of times greater than the distance across our solar system, and at currently achievable speeds, a trip to the closest star would take in excess of 10,000 years, one way! Even at speeds close to the speed of light, to fully explore even the nearest star systems for life would take many centuries. (This is why Ripley "hibernates" in the *Alien* movies.)

And there are at least 100 billion stars in our galaxy alone, while the Milky Way is just one of over 100 billion galaxies in the observable universe.

At the very least, it is clear that to travel beyond our solar system on a human timescale, we would have to reach speeds much greater than are possible now. To appreciate the magnitude of the problem, let's consider the onboard fuel requirements just to accelerate a spacecraft to 25 percent the speed of light, at which rate a one-way trip to Alpha Centauri would take a mere 10 years. (I will ignore here the fact that it would take at least a year to do the accelerating, if you want to survive that process.) If we make use of the rocket equation—which, I remind you, is an underestimate of the fuel requirements—and specify conventional rocket fuel, the fuel mass would be 1 followed by approximately 20,000 zeros times the payload mass! To deliver just a single atom to the nearest star would thus require more fuel than is available from all the matter known in the universe! I think even Congress would realize that this is not the way to go.

Conventional fuel is obviously the straw man in this conundrum, though. No one has seriously suggested trying to reach the stars with rockets of the type we use to orbit the Earth. But there is a lot of creative thinking going on in this business, so a number of people are claiming that at least a manned Mars mission in the next decade or two is a reasonable goal. From there, who knows? The exigencies of space travel require us (and any aliens out there who want to get here, however advanced their technology) to utilize two simple ideas, both of which, oddly, hark back to the prescient 1960s: *Small Is Beautiful* and *Live Off the Land.*

CHAPTER FOUR

A Cosmic Game of Golf

SCULLY: Why is it so dark in here?
MULDER: Because the lights are out.

Whether you are Han Solo, Jean-Luc Picard, or some slimy alien, the most daunting challenge you face once you engage your thrusters is not how to zip around the sky with the ease of a hummingbird. It is to get moving in the first place.

Size isn't important, some of us are often told, but as consoling as that might be in certain circumstances, it isn't true when it comes to building spacecraft. Say you are stuck on some very slippery ice. The only way to get back to shore is to use propulsion. You could push yourself off a nearby rock very effectively. In this case, a lot of mass (the rock) moves away slowly when you push it, and you move away very fast. However, if you have to bring your own fuel with you, most people would not opt to carry a rock along for such a purpose. Instead, you might pack your backpack more lightly, with a golf club and golf balls. When you want to move on the ice, you unpack the balls, place them on the ice, and hit them away one at a time. Each ball you hit, being light, doesn't cause you to recoil rapidly. But since you can propel each golf ball much faster than you can throw a heavy rock, and since what matters for the outgoing propellant is a combination of mass *and* velocity, by the time you've exhausted your supply of golf balls you may be moving much faster than you would have if you'd simply thrown a rock.

The process just described is counterintuitive—probably because most people (and I'm no exception) prefer instant gratification. But Aesop was right—slow and steady wins the race, if "slow and steady" refers to acceleration and the race is long enough. If achieving a greater final cruising speed is more important than reaching your cruising speed quickly—which is certainly the case for space travel, as opposed to (for example) the firing of ground-to-air missiles—what you need more of is not *thrust,* the quick acceleration you get from expelling a lot of fairly slow-moving material, but *impulse,* the large final velocity achieved by the steady release of small amounts of material that travel away very fast. (Given the rapid achievement of cruising velocity after Jean-Luc Picard engages the impulse drive aboard the *Enterprise*, it should probably be called the thrust drive.)

What are the existing and potential rocket-engine equivalents of "golf balls" currently under investigation by researchers? Clearly what is wanted are light projectiles and the energy to send them out very fast. The lightest atom in existence is hydrogen (1 proton plus 1 electron), so that's the obvious projectile of choice, and essentially all existing proposals use it as the propellant. The source of energy to boost the propellant depends on available technology. The most likely candidates in the short term are nuclear thermal rockets, currently being studied by a group at NASA's Lewis Research Center, near where I live in Cleveland, Ohio. Here, a nuclear reactor simply heats a liquid—for example, liquid hydrogen—to as high a temperature as one safely can (about 2,500°C, given current materials) and then spurts the "steam" out the back end of the reactor. In this way, experimenters have achieved exhaust velocities as great as 10 kilometers per second, or 1/300 of a percent of the speed of light.

While this velocity may not sound useful as far as interstellar travel is concerned, it would be a boon to interplanetary travel. Using this technology, one might send a craft to Mars and back with only about 5 or 10 times the mass of the payload in fuel. The problem is that the nuclear-reactor technology required is large-scale and currently politically incorrect. However, if the political climate changes and the use of nuclear reactors in space becomes acceptable, a better option might be what is called a nuclear electric propulsion system. In this system, instead of using nuclear power to heat up gas as a propellant, one could use the nuclear reactor to generate heat which is then used to generate electricity, as we now do on Earth. One could then use large electric fields to accelerate atomic nuclei (such as the proton that forms the hydrogen nucleus), much as we now do in the large elementary-particle accelerators we've built to study the fundamental structure of matter. In the versions that have been explored, these charged particles would fly out the back of the engine at close to 50 kilometers per second. (Modern particle accelerators bring particles almost to the speed of light, but these devices are generally many miles long and at present require power sources much greater than would be available to a spacecraft.) In this way, one might build rockets for interplanetary travel which required only 2 to 4 times the mass of the payload in fuel, and, more important, one could build up speeds that allowed travel even to the outer planets in months instead of years.

None of these technologies, however, would help much for the near-light-speed travel required to reach the stars. If one wanted to achieve these speeds by internal propulsion, one would need a propellant that traveled at a significant fraction of the speed of light.

Enter *Star Trek*. The *Enterprise*'s impulse drive just referred to, used for sub-light-speed travel, is powered by nuclear fusion. This is precisely the method one would choose in the real world to approach the speed of light. When one combines, or "fuses," hydrogen nuclei—or the nuclei of its heavy cousin, deuterium—to form helium, the energy released is sufficiently great that the helium nuclei receive kicks that propel them to about 5 percent of the speed of light. One can imagine, at least in principle, accelerating spacecraft to 3 or 4 times this value, with the fuel weighing less than 100 times the mass of the rest of the payload—a quantity that of course is still nothing to shake a stick at.

Even here, success breeds its own problems. The very reason that fusion is so effective—the great amount of energy released—also presents a concern. How do you ensure that all the energy gets out the back and doesn't melt the engine—or the spacecraft, for that matter? Remember how much heat the saucers in *Independence Day* would have to generate! Of course, as we physicists like to say, this is an engineering problem—so let's not worry about it now.

But there are other problems. If you want to stop as well as start, you have to bring along the fuel required to stop, which will be the same amount as was required to start. This means you need to bring not just nearly 100 times the mass of the rest of the payload in fuel, but 100 times the mass of *both* the payload plus fuel required to stop, or about 10,000 times the mass of the original payload. And that's just to start and stop! It doesn't include the fuel required for the return journey.

This might suggest that we should not fool around with any propellant that travels slower than the speed of light itself, if we want to head out at near light speed. So, in this case, why not just send out a beam of light behind you in order to accelerate? The problem here is that visible light carries with it an extraordinarily small amount of energy. Thus, while in principle one could eventually accelerate to light speed by shooting a laser beam out behind the ship, it would take an incredibly long time to do so.

Let's get back to being stuck on the ice: Emitting visible light is analagous to throwing off grains of rice. This will eventually get you to shore, but it will take an awful lot of rice to start you moving. Accelerating and decelerating single atoms with visible light from lasers has been demonstrated in the laboratory, and it works very well. Unfortunately, we and our ships are made of an awful lot of single atoms.

We shouldn't despair, however, because there is a way of producing radiation at a much higher energy than the type we normally produce with lasers. If we take a particle of matter and annihilate it with a particle of antimatter, the products can shoot out at near the speed of light, and, more important, they will carry away all the original energy of the matter-antimatter pair that were annihilated—just as if you had kicked out the original particles at something like the speed of light. This is a process tailor-made for rocket propulsion. As bomb makers like to say, there is no bigger bang for your buck.

Antimatter may sound like pure science fiction, and indeed it is central to *Star Trek* propulsion. But it isn't science fiction at all. A brief review is in order: The existence of antimatter in nature is an inevitable consequence of the theory of relativity, combined with the laws of quantum mechanics governing the nature of subatomic particles, about which I shall have much to say later. Antiparticles are identical to their normal-particle partners in all respects—same mass, spin, and so on—but they have the opposite electric charge, and a few other, more obscure characteristics are reversed as well. When a particle and antiparticle interact, they may annihilate into pure radiation, which carries away the energy stored in their mass. By 1930 it was understood that antimatter should exist, and (coincidentally) within 2 years the antiparticle of the electron was discovered amid the cosmic ray particles that shower the Earth every day from outer space.

Now, the problem is that antimatter is generally hard to come by. Not much of it exists in the universe, although a discovery at the time of this writing suggests that there is a handy fountain of antimatter at the center of our galaxy, if only we could get there to tap into it. (The only practical way to get there to tap into it might be in light-speed ships powered by antimatter.) I described at great length in my last book how we actually produce charged particles of antimatter on Earth in our large particle accelerators. I now have a chance to update those discussions by describing several recent experimental discoveries. While I predicted at the time that such results would be produced in this decade, I was completely wrong about where they would occur.

The big news in 1996 was the production, for the first time, of neutral antiatoms at the CERN (European Center for Nuclear Research) laboratory in Geneva. Antiatoms are the experimental holy grail in antimatter research. The antiparticles we normally produce in our laboratories are the antiparticles of elementary particles—protons, say, or electrons—not the atoms composed of combinations of these particles. These electrically charged antiparticles are not particularly exotic, other than having the opposite charge of their particle partners. But atoms are electrically neutral (which is a good thing, because if they weren't, the electric forces between each of us would be strong enough to crush us completely, or perhaps first cause each of us to explode).

There is a fundamental theorem in physics that if the universe were made of antimatter instead of matter, the antimatter would have combined to form antiatoms that would behave essentially like normal atoms. Antistars would look like normal stars—they would emit the same frequencies of light, and so forth—and antibaseballs would fall toward anti-Earth at the same rate as normal baseballs fall to Earth. Anti–Captain Kirks would not be evil, and anti–Mr. Spocks would not laugh or grow beards. The universe would look essentially the same. While we have no reason to doubt this based on every indirect test we have done to date, there is one direct test we have not yet performed. Until 1996, we had never observed antiprotons combine with antielectrons to make antiatoms. Until we had antiatoms, we could never do any experiments on them directly to see if they behave like normal stuff.

The first antiatoms created at CERN were produced in an exotic way. When very-high-energy particles collide with a thin target, all sorts of new particles are created, which stream out behind the target in the same direction as the original particle beam. If you take a very-high-energy beam of antiprotons (used at CERN to explore the fundamental structure of matter) and bombard a target with this beam, some of the new particles produced are positrons, the antiparticles of electrons. Very rarely, a positron may be kicked out behind the target with a velocity close to that of remnant antiprotons in the beam which find their way through the target. These positrons will travel alongside the antiprotons, and if you're lucky, a positron and an antiproton will combine to form antihydrogen.

At the CERN accelerator, the antihydrogen atoms produced this way were traveling at close to the speed of light, and were disrupted within several millionths of a second. This time period was far too brief for accurate experiments to be done with these antiatoms. What one would like to do is produce antihydrogen at rest in the laboratory and keep it around for an appreciable length of time. One could then see, for example, whether or not antihydrogen atoms fell in the Earth's gravitational field at exactly the same rate as hydrogen atoms. Or one could excite the antihydrogen atoms with an electric current to see whether or not they responded by emitting light with exactly the same frequencies that hydrogen does.

This is precisely what experimentalists at CERN are trying to do. Money has just been appropriated to build an antiproton decelerator, which will slow down antiprotons produced in CERN's synchrotron so that they can be trapped, cooled further, and combined with stored positrons in the lab to make small amounts of stable antihydrogen.

But how, you may ask, can antihydrogen be stored in order to be experimented on? After all, if the antihydrogen atoms hit the walls of a container made of normal matter, they can annihilate with the protons and electrons in the walls. Charged antimatter is simple to confine in a box without having it come in contact with the walls. Charged particles (antiprotons, say) move in circles when put in a magnetic field, so they can be confined in what is called a magnetic bottle—a torus, or doughnut-shaped ring, with a magnetic field inside that keeps the particles traveling in a circle around the center of the torus, away from the walls. This is how we store antimatter particles in particle accelerators at the present time, and in my last book I criticized the *Star Trek* writers for stowing the *Enterprise*'s antimatter aboard ship in the form of heavy-hydrogen antiatoms (antideuterium) instead of as charged antiprotons, which would be a lot easier to store.

I now think I was a little too hard on the writers. One reason you would want to store antimatter fuel as atoms instead of as independent sets of positively and negatively charged particles is the same reason CERN is trying to cool and store neutral antiatoms. If you eventually want to build up a large amount of material—be it thousands of antiatoms, as at CERN, or billions of billions of billions of antiatoms, as in a starship—you can't continue to work with charged particles. The electrical repulsion between charged objects is so unimaginably great that it is virtually impossible to store large amounts of them at any reasonable density. In fact, if the Earth contained, on average, one additional electron per 5 billion tons of material, the force of repulsion on an electron at the Earth's surface would counterbalance the gravitational force holding it down. A greater proportional increase and the Earth would blow itself apart!

Well . . . so how can you trap and store neutral antiatoms? You use magnetism again, but this time in a trickier way. The nucleus of an antihydrogen atom consists of a single antiproton. Since one of the fundamental properties of antiprotons is that (like protons) they have a property known as nuclear spin, they act like little magnets. In a strong external magnetic field, they spin in such a way that their own internal magnetic fields tend to line up with the external field—because it would take more energy for them to line up in, say, the opposite direction. Now, if you create antihydrogen atoms and then cool them down to a few thousandths of a degree above absolute zero—which, remarkably, we can do nowadays—then essentially none of the antiatoms will have enough energy to line up their nuclear magnetic fields in the opposite direction to the external magnetic field. Imagine then that we have a bunch of very cold antiatoms with their spins all lined up in some direction, and we place them in an enclosure with a strong magnetic field around the outside aligned in the opposite direction. If the atoms are cold enough, none of them will have sufficient energy to hang out in the region of a strong magnetic field, so they will tend to cluster at the center. You will have a magnetic trap.

Magnetic traps have already been used successfully to confine normal atoms, and the principle should work exactly as well for antiatoms, once we create them. Such a program is planned to be in operation at CERN by 1999. The antiproton decelerator's estimated cost is about $5 million, and it should allow storage and detection of about 1,000 antiatoms per hour. That's about 9 million antiatoms per year; at that rate it would take somewhat more than a million times the present age of the universe to make enough antiatoms to propel a flea to near the speed of light.

So antimatter propulsion is not practical right now. But one day, if we, or any other creatures, want to travel at near light speed and carry enough fuel along to do it, antimatter is the best and perhaps only way to go. Even here though, there are huge problems: To do a round-trip this way would require 16 times the ship's mass in antimatter! Carrying and containing antimatter 16 times the mass of a large spacecraft for a period of, say, 20 years—probably the minimum amount of time needed for a round-trip to the nearest star system outside our own—is a logistical nightmare. Even the *U.S.S. Enterprise* has antimatter containment problems on a more frequent basis than that. There has to be a better way!

CHAPTER FIVE

There, and Back Again?

> Einstein had a little theory.
> It had something to do with relativity.
> Well, Einstein put that theory to the test,
> That's why he looks confused, and his hair's a
> mess.
>
> —*New Rhythm and Blues Quartet (NRBQ)*

Practicality is something we often dispense with when it comes to imagining the future. Part of the fun of physics, and science fiction, is recognizing that to make any progress in the world we can't limit ourselves to thinking about what we're capable of today. But at the same time we need to keep in mind that whatever people, or aliens, do build will have to be practical in its own time. If we want to speculate on what form interstellar travel (or any future technology, for that matter) will eventually take, we have to try and imagine what will be easiest—given the laws of physics we already understand, along with the possibilities they don't yet rule out.

Necessity is always the mother of invention, both in the real world and the world of science fiction, even when what is contemplated appears implausibly extreme. As Jean-Luc Picard once announced to Data, reprising a remark made generations earlier by Captain Kirk to McCoy, "Things are only impossible until they're not!" If round-trip rocket propulsion seems the most preposterous way of traveling to the stars, we must be willing to consider alternatives, even those that initially seem absurd.

And we are doing this already, because the same problems that will confront future interstellar travelers are confronting, on a smaller scale, today's engineers, as they wrestle with the problems of manned round-trip travel through our own solar system. They may well be thinking about how their predecessors handled the same problem. Columbus didn't need fuel to set sail across the Atlantic; he used the wind. Lewis and Clark didn't bring along the fuel they needed to explore the innards of North America; they hunted and fished as they went along. The lesson is clear. If you want to explore strange new worlds where no one has gone before, you probably have to live off the land to do it.

A local version of this strategy has been proposed by rocket engineer Robert Zubrin as a way of getting to Mars at a cost we may be able to afford. It is known as the Mars Direct approach, and it calls for sending astronauts to Mars in a craft containing only the fuel needed for the outbound voyage. Fuel for the return trip could be manufactured on the Martian surface, using very simple technology—technology so simple, in fact, that Zubrin, who is not a chemical engineer, has built a working prototype on Earth.

The Martian atmosphere is 95 percent carbon dioxide, and atmospheric CO_2 can easily be filtered out, pressurized, and stored as a liquid at Martian surface temperatures. Zubrin's proposal involves bringing along a small amount of hydrogen and reacting it with the carbon dioxide to produce methane and water. Since this reaction is exothermic—that is, heat-releasing—

it requires no input of energy to drive it; rather, it occurs spontaneously in the presence of a catalyst made of nickel or ruthenium. The methane and water are easily separated, and the water is then split by electrolysis into hydrogen and oxygen. The hydrogen is recycled, and the oxygen is refrigerated and stored. When it's time to go home, just mix the oxygen and the methane, and you will have produced a high-performance fuel in a form that can be stored for a long period of time.

It might be argued that sending people to Mars without fuel for the return journey would not provide the kind of safety margin that keeps NASA in the business of manned spaceflight. Zubrin's ingenious answer to this is to send a spacecraft containing the fuel fabrication facility to Mars *in advance* of the manned spacecraft. Only after this automated craft had safely landed and produced the requisite fuel would the manned mission be launched.

Finally, the question arises of how to transfer the fuel from the fabrication facility to the return vehicle. The answer is that it is simpler to transfer the astronauts. The original landing vehicle containing the fuel facility will *become* the return vehicle to Earth. The astronauts will land in their module, and when their time on Mars is through they will transfer on the Martian surface to the fully fueled return vehicle. The combined spacecraft will in the meantime serve as the Mars base, housing the astronauts for up to 2 years, until an accessible return trajectory is available.

Of course, there are a plethora of other concerns to worry about: radiation exposure during the Mars round-trip and power on the Martian surface, artificial gravity during the months in transit so that the astronauts' muscles don't atrophy, and so on, but these are solvable in principle once one knows that one can send a crew to Mars and back with sufficient fuel for a reasonable amount of money. Depending on the size of the crew and the necessary radiation shielding, cost estimates for the round-trip are in the neighborhood of $10 billion to $50 billion. Not exactly cheap but realizable—comparable in 1960 dollars to the money spent to send men to the Moon.

How might one then adapt this idea for travel to the stars? Can we assume that before we're visited by aliens, we should expect a large ship to land on Earth—in the Mojave, or somewhere outside Las Vegas, or near Roswell, say—and start producing fuel? I don't think so.

We know a tremendous amount about Mars, but if we are traveling to another star's planetary system, we probably won't know enough details about where we're going to send an advance hospitality vehicle there to bring us home. I know of no such precedent in human exploration. However, various optimistic individuals have proposed that instead of powering a spacecraft using fuel obtained either on Earth or at the destination, one should do as the earliest explorers did and harvest fuel along the way.

Now, the density of matter in our region of the galaxy is very small—about 1 proton (on average) per cubic centimeter. This makes scooping up matter, or antimatter, for use as fuel impractical. However, the universe is also full of radiation. The first person to propose the use of radiation power was also the first person known to have written a science fiction story involving space travel. Johannes Kepler, the discoverer of the laws of planetary motion, was a busy fellow, with a life full of interruptions. In between his contributions to science, he successfully defended his mother against the charge of witchcraft and wrote a story about traveling to the Moon and back. He also observed something just as timely, given the recent return of comet Hale-Bopp: Whether comets are traveling toward the Sun or away from it, their tails point away from it; hence, the Sun must be exerting a kind of pressure. This prompted Kepler, in 1609, to suggest that we would one day design ships that could sail on these "heavenly breezes."

There is indeed a solar wind, a stream of charged particles

moving out from the Sun into space at high velocity. However, this velocity is still only about $^1/_{10}$ of 1 percent of the speed of light. While a solar-wind-powered sailing vessel, coasting along on the initial push it got from the solar wind, might be useful for interplanetary travel, it would not be particularly useful for interstellar travel—at least on a human timescale.

In addition to the solar wind, sunlight itself produces a pressure—any sort of light produces a pressure. But this pressure is very small. After all, if the Sun's photon pressure packed a wallop, the Earth would be pushed around by it. Nevertheless, certain bold futurists have suggested using solar sails to carry us to the stars. To get enough propulsion to accelerate a 1,000-ton spacecraft even to 10 percent of the speed of light in a year would require a solar sail perhaps 100 miles across, and in order that it not weigh more than the spacecraft it would have to be less than $^1/_{1000}$ the thickness of a kitchen garbage bag.

Others have suggested improving upon the Sun. While the Sun is very bright, it shines in all directions. Think of all the sunlight wasted that way! Why not build a powerful space-based laser, perhaps powered by the Sun, which would direct a concentrated beam of light toward a sail big enough to encompass the beam even at vast distances—perhaps $^1/_4$ the width of Texas? Several years before the spacecraft reached its destination, another beam could be turned on which would reach the spacecraft in time to decelerate it, using a series of reflectors.

All these ideas have their own problems, of course—some involve open questions of fundamental principle and others involve specific engineering issues. They all also require tremendous resources to build the huge sails and the lasers. And they depend for their success on specific interstellar conditions. Just as one must take wind variations into account in a sailboat, navigating the interstellar winds would be a difficult business. Similarly, one cannot travel under external laser power if one is not within the laser's sights. Finally, none of these methods allow for unscheduled stops; missions would have to be planned completely in advance, and the discovery of something interesting along the way would probably have to be recorded for the next mission to explore.

Now, everything I have talked about thus far, even fusion and antimatter drives, involves "conventional," well-understood physics. I think I have adequately demonstrated that any aliens who want to get here probably can't resort to such conventional physics. But who expected them to? As Mulder has noted, rather wistfully, to Scully, "When conventional science offers no answers, may we finally turn to the fantastic as a possibility?" My answer is, "Yes, as long as the fantastic isn't impossible!"

OK, then, what about warp drive, wormholes, antigravity, and all the wonderful exciting unknowns associated with the nature of spacetime? One could write a whole book about what may be possible, but that has already been done. I want here to set out what might be possible in practice, as opposed to what might be possible in principle. The wonders of general relativity allow all sorts of incredible things to exist in principle, from warp drive to time travel. That alone warrants thinking about them, and writing about them, and I even spend some of my own research time trying to make some progress in this regard. But we began here by asking how the spaceships that might one day actually be built might behave.

Here is the place to state unequivocally that I think these things will *never* be practical for real space travel, although they may well be possible in principle. Even glimmers of hope can become blindingly bright when people are intent on maintaining any hope at all. When I and others began popularizing the idea that it's still an open question whether or not warp drive and time travel are possible in our universe, I was amazed at the excitement and speed with which this idea propagated—not just among the fans of popular science but throughout the academic community. Even NASA seemed to be listening, and has invited me to speak at a symposium on nonpropulsive methods of space travel, including warp drive and wormholes.

The fundamental and formidable energy problems that have thus far kept human beings away from even the closest planets pale in comparison to those that arise when you turn from conventional Newtonian propulsion to the fantastic possibilities opened up to us by Einstein. Let me remind you of some of these, so that I can then tell you about some recent exciting new discoveries, and also reveal a little secret about warp drive which I don't think has been discussed in print before.

By now, it's clear what a Herculean task it is to imagine traveling at near light speed through space in realizable rocket ships. But why bother traveling through space if you can make space do the traveling for you? Einstein's general theory of relativity tells us that space itself responds to the presence of matter by expanding, contracting, and bending. If this is possible, then a brave new world of "designer universes" opens before us.

Within the context of general relativity, you don't have to move at all to travel throughout the universe. You can move at the speed of light and yet be sitting still. In fact, you are doing that right now as you read these words. While you and I are more or less at rest with respect to each other and to our nearby surroundings, we are traveling at the speed of light relative to a being in a galaxy at the other end of the visible universe reading the Klingon translation of this book. And that being is also at rest with respect to its local surroundings, yet it is traveling away from us at the speed of light.

How can we be both traveling and sitting still? Simple: The *space* between us is expanding.

This idea is what validates warp travel. One can show explicitly in the context of general relativity that the following is possible in principle: Say you want to travel to the nearest star but don't relish spending 10,000 years in a rocket ship. Well then, all you have to do is travel a little more than 3/4 of the way to the Moon, to the point where the Moon's gravitational pull balances that of the Earth, and you can remain at rest there with your engines off. Now arrange for the space between you and the nearest star—all 4 light-years worth of it—to collapse in, say, 1 second, while the space between you and the Earth, formerly only about 180,000 miles, now expands correspondingly in the same short period of time. After space has done its thing, you look around and find that you're now only 180,000 miles away from Alpha Centauri and some 4 light-years away from Earth—all without moving! Then simply turn on your engines and go the rest of the way.

As fishy as this may sound, the equations of general relativity have been solved to reveal exactly such a possibility. I don't want to downplay how truly remarkable this is. In fact, more or less the same physics might make even stranger phenomena—traversable wormholes, say, and time machines—possible in principle. However, consider the following:

1: Expanding space would require a kind of matter unlike anything we have observed directly—a kind of matter that is gravitationally repulsive rather than attractive. While the laws governing the behavior of matter on the subatomic scale make such a phenomenon realizable on that scale, we have no idea whether such material could be created *even in principle* on a macroscopic scale. Preliminary indications are not particularly encouraging.

2: It would take more energy than the Sun will emit in its entire lifetime to make such material useful for moving any macroscopic object, even if the material could be produced.

Now for the new results: In the past year, a number of researchers have subjected warp-drive theory to the same kind of scrutiny previously applied to the idea that wormholes might be used as shortcuts through space. Their findings have been no more encouraging. The theoretical physicist Larry Ford and his colleagues at Tufts University have shown that in order not to violate known laws associated with energy conservation, space must expand and contract (at any one time) only in the thin surface layer of a bubble surrounding the spacecraft. It turns out that to maintain a region of exotic matter within a thin shell encircling a macroscopic object like a spacecraft, you'd need an energy roughly equivalent to 10 billion times the entire mass of the visible universe! Perhaps we might imagine transporting single atoms at warp speed, but not spacecraft. The same kind of energetic arguments apply to wormholes and therefore to the time machines that might make use of them.

There is a much more worrisome problem of principle, which makes warp drive look even less likely, if such a thing is possible. I did not explicitly mention it in my earlier book, because I believed that the other problems were bad enough. Perhaps I should have known better. It's this: While warp drive allows you to travel globally from one point to a distant point faster than the speed of light, it still won't get you there any sooner. How so? Well, say you want to travel 1,000 light-years in 1 second using warp drive. In order to arrange for the space in front of you to collapse, you must arrange for a proper configuration of matter to be distributed throughout it. To do this, you must, at the very least, send a signal all the way throughout this space. But it takes at least 1,000 years for this signal to spread across the region. Thus, while you could (in principle) travel arbitrarily fast once the warp front started to collapse, the "countdown" to takeoff would last 1,000 years. I suppose there is some comfort in being able to spend the 1,000 years waiting in comfortable surroundings instead of sitting inside a cramped spacecraft, but the end result is the same. However you slice it, you can never get from "here" to "there" in less than 1,000 years from the time you first start trying. As wonderful as the possibility seems, in the end warp drive is a cosmic letdown. Take that, Fox Mulder!

Energy is energy, and even a million years from now, when we will know a lot more physics than we do today, the energy requirements to transport us throughout the galaxy will still be the same, and the energy required to manipulate gravity to bend it to our will seems to be greater than all the energy in the galaxy put together. This is why most physicists, myself included, find it so unlikely that Earth has been visited by aliens, especially aliens from a sufficiently advanced civilization willing to exhaust the necessary resources to travel all the way here just to insert metal objects up people's noses or abduct the patients of a Harvard psychiatrist. Even if they did want to do kinky experiments, it hardly seems worth it.

Fox Mulder, who has certainly replaced *Star Trek's* Q as the most quotable person on television, once argued that "the easiest explanation is also the most implausible." For many people, the easiest explanation for the vast number and variety of tales of alien sightings and abductions is that aliens have been here. But to physicists, this explanation is the most implausible—simply because the more mundane possibilities involve requirements far less daunting than those that face any interstellar travelers.

Because we seem to be forbidden by energetics (if, perhaps, not by physics) from traveling at speeds greater than light, the plausibility of Area 51, Roswell, alien implants, and all of that becomes even more remote. Why should aliens devote the necessary resources to visit us if they are not aware that intelligent life exists on Earth? But in order to be aware of this, they would have to have received signals of our existence. We have only been emitting such detectable electromagnetic signals—via *I Love Lucy, Star Trek, The Twilight Zone, NBC Nightly News,* and so on, for little more than half a century. By 1947, the year of the first flying saucer sightings and of the Roswell incident, our broadcasts would just have begun to reach the closest star systems. It seems wildly unlikely that any civilizations living there would have had time, even had they possessed the necessary resources, to launch a mission to Earth that would arrive by 1947. The aliens in the movie *Contact*, who detected the TV signal showing Hitler opening the 1936 Olympics, sent back a reply which didn't get here until 1996.

There is a loophole I haven't yet discussed, and it is one that is often brought up at my talks, either in the context of alien visitation or of our current view of the universe. What if the laws of physics aren't the same *out there* as they are here! Indeed, if Q can transcend our laws of physics, why can't the universe? One often finds in science fiction stories that in certain "weird places" the laws of physics don't behave as they are supposed to. I still vividly remember the terror I felt as a child, watching a *Twilight Zone* episode in which the walls of a house suddenly became incorporeal gateways to another dimension and a small boy about my age fell through.

It is impossible to guarantee—at least until one has turned over every last rock and explored every last nook and cranny—that there are no *Twilight Zone*s in the real universe. So why are we physicists so conceited as to assume that our laws are universal? "What cosmic gall!" my wife often exclaims when confronted with like assumptions.

Well, there are two answers, but they are both essentially the same. The first is that 400 years of success has indeed made physicists conceited. The second is that in those 400 years of success, every test we have performed to check for the universality of physical laws has come up positive.

Rather than dwell on the details of the history of physics, I want to tell you about a modern discovery, which I believe convincingly underscores the universality of the fundamental laws of physics as we know them.

When confronted with questions about the universality of

physical laws, scientists usually turn to the stars. The editors of *Social Choice* and other postmodern journals may suggest that the laws of physics would be different had they not been developed by dead white males, but I find comfort in objective reality when I look at the sky on a starry night. Around distant stars there may be planets where female symbionts inhabit and govern otherwise male bodies and minds, like the Trill in *Deep Space Nine,* but even there the laws of physics they develop will have to account for the fact that their Sun (or Daughter?) shines with exactly the same colors as ours. There is nothing more telltale—not even a fingerprint, or a DNA trace—than the spectrum of light emitted by an object when it is heated up. Every element emits its own unique combination of colors, and it was one of the great successes of twentieth-century physics to catalogue those spectral features that had already been observed and then to predict those that hadn't. The fact that distant stars shine with the same set of colors emitted by hydrogen gas when it is heated in a lamp in a terrestrial laboratory not only tells us that the stars are made mostly of hydrogen, it also tells us that the very laws of electricity and magnetism, which (together with the laws of quantum mechanics) produce these spectra, must be the same there as here.

So much for the stars, but what about the space *between* the stars. What about the universe itself? Well, thanks to NASA, we now have compelling direct evidence that the fundamental laws of physics as we know them apply on the scale of the entire visible universe and, moreover, have so applied for most of its lifetime. You may have heard of the Cosmic Background Explorer (COBE) satellite, which was launched in 1989 by NASA to measure the properties of the radiation left over from the Big Bang from which our universe emerged some 10 to 15 billion years ago. COBE famously succeeded in finding minuscule fluctuations in this so-called cosmic background radiation, constituting the "seeds" of the cosmic structures we observe today, but first it measured the spectrum of the primordial radiation and confirmed what had been predicted—that it was of the form known as blackbody radiation.

All you really need to know at this juncture about the blackbody spectrum—so named because it is the spectrum emitted by a perfectly black object when it's heated—is that the correct understanding of it was the driving force behind almost all the major results of twentieth-century physics. The investigation of blackbody radiation led to the development of quantum mechanics and the correct quantum-mechanical treatment of electricity and magnetism. More important, at the very basis of the blackbody spectrum is a profound and subtle understanding of the statistical behavior of myriad individual particles. This understanding, called statistical mechanics, is at the heart of almost every calculation done by theoretical physicists today and almost all observed phenomena. It was developed to explain why you can't tell by looking at, say, a movie showing the collision of 2 billiard balls whether the film is running backward or forward, whereas when you observe the collisions of 16 billiard balls—as when the cue ball hits a freshly racked set—that symmetry is lost. The principles involved in statistical mechanics are so subtle that two of its developers killed themselves because of the initial resistance to their ideas.

In any case, it turns out that the cosmic background radiation left over from the Big Bang not only exhibits a blackbody spectrum but it is the most perfect blackbody spectrum ever measured—closer to the theoretical prediction than anything we have been able to create in the laboratory. We can therefore use the universe itself to test the predictions of quantum mechanics. To turn it around for the argument at hand, we now know that even the most subtle and complex laws that lie at the foundation of modern terrestrial physics apply to a radiation bath that permeates the entire visible cosmos, in space and time. It would be hard to ask for a better reason to believe that if there are *Twilight Zone*s they are well hidden, and therefore probably irrelevant.

■ ■ ■

In spite of all of this—in spite of the evident impossibility of launching realistic spacecraft to make round-trip visits to other stars, the implausibility of alien visitation, and the universality of the roadblocks and effective speed limits marking interstellar travel—I am firmly convinced that our destiny does lie in the stars. We will, one day, travel beyond our solar system. How can I say this with a straight face, after all that I have argued here? Well, every obstacle I have described lies only in the way of making a *round-trip,* on a *human timescale.* But the key to our future in the stars is that neither of these conditions need be fulfilled when we do venture out into the galaxy, as I believe we one day must.

CHAPTER SIX

Seeing Is Believing

> I believe I am the most fortunate sentient in this sector of the galaxy.
>
> —*Data*

Fifty years ago, Martians were the prototypical extraterrestrials, with Venusians a close second. As we learned more and more about our solar system, however, our expectations for finding life (let alone intelligent life) lurking on Mars or Venus began to fade. With the exception of a few Hollywood stars, the rest of us accepted the fact that we lived on the only planet orbiting our Sun which had ever harbored intelligent life.

How much has changed in the past year! The claim by a team of NASA and university researchers that a meteorite from Mars known as ALH84001, which fell to Earth some 13,000 years ago and was later discovered in Antarctica, showed fossil evidence of microscopic life-forms electrified the world. Perhaps the rest of the solar system is not barren after all.

The search for life on Mars has its roots in the search for the origins of life on Earth. Until perhaps a decade ago, it was felt that in order to flourish, organic life, like Goldilocks, needed conditions that were "just right"—enough water, warmth, and light, but not too much. But scientists exploring remote and inhospitable locations ranging from boiling vents in the deep seafloor to the frigid, wind-scoured valleys of Antarctica, from the burning sands of the Gobi desert to the sulfuric ooze from oil wells, have discovered that various forms of primitive life (like various forms of less primitive life such as race-car drivers and mountain climbers) choose to live on the edge. These extremophiles, as they are called, exist in all the wrong places. They may not flourish (indeed, some of them just barely hang on), but they survive, sometimes without light, heat, oxygen, or water—all the standard ingredients we once thought necessary to life.

The evidence in favor of possible primeval life on Mars is controversial, but it does point to several interesting possibilities. The fossilized microbes that investigators claim to have found in ALH84001 would be more than 3 billion years old. They date to a time when the Martian surface was warmer and wetter, and thus more hospitable to life. Why Mars became barren and Earth did not is not fully understood. However, perhaps what is most significant about the claimed discovery is that the discoverers did not have to go to Mars to find the rock; it was sitting there, waiting to be found, on the windy ice-covered plains of Antarctica.

Equally significant, perhaps, is the fact that this same location is where scientists have discovered primitive terrestrial life-forms called cryptoendoliths living inside frozen rocks. And deep underground in the frozen permafrost of Antarctica and Siberia, microbes have been discovered in various states of activity, some of them having lain virtually dormant for over 3 million years.

It is now clear that meteor, comet, and asteroid impacts on planetary surfaces impart enough energy to kick projectiles into interplanetary space. This means that the Earth is not a closed ecosystem! If matter is exchanged between planets, then certainly organic materials might be, too—including, perhaps, primitive self-reproducing life-forms. (It is highly unlikely that any advanced form of life would survive the catastrophic ejection and subsequent interplanetary voyage.) Moreover, if primitive life-forms can remain dormant for millions of years until conditions are appropriate for them to "turn on," then it's perhaps possible that life on one planet could seed life on another.

This is reminiscent of the panspermic theory that Francis Crick proposed, not altogether facetiously, a while back. Similar ideas have been proposed in science fiction novels and movies, with the source of the "seeds" usually being alien intelligences who later return to see how their offspring are doing. In a particularly creative use of this idea, the writers of *Star Trek* were able to explain why a great many of the extraterrestrials the *Enterprise* crew encounters are humanoid. Jean-Luc Picard, carrying on the work of the archaeologist Richard Galen, discovered that the primordial seas of many different planets had been seeded with DNA provided by a long-dead civilization.

In any case, the discovery of what might be fossil evidence of Martian life, combined with interplanetary transport afforded by cataclysmic planetary collisions, suggests that the discovery of extraterrestrial life in our solar system may in fact be nothing of the sort. Who is to say that such life will be unrelated to our own? We may discover only our distant cousins! In fact, it appears that nonintelligent life-forms can survive processes that eventually lead to the demise of their home planets. The frozen bacteria in the Siberian permafrost demonstrate that primitive life-forms are capable of outlasting devastating climatic change. Could such microbial life survive long enough to be ejected and seed another world?

The notion that life on Earth may well not have originated on this planet received further support from observations of the Hale-Bopp comet. (No, I am not referring to the observation of an alien spacecraft carrying members of our mother civilization!) Spectroscopic data indicate the presence of over 100 different types of relatively complicated organic molecules on the comet, including glycine, an amino acid. It has been argued that enough organic material—and water, as well—could have been delivered to the Earth's surface by cometary impacts during its early history to provide the wherewithal for all organic life on our planet. This is supported by recent evidence that the Earth is being continually bombarded by up to 30 small, water-bearing comets per minute, and by observations of the impact of comet Shoemaker-Levy on Jupiter, which indicate that some water from the comet made it down into the planet's atmosphere. Perhaps—in a classic reversal of the typical scenario wherein we colonize the solar system—the solar system colonized us.

This colonization might explain the relatively early appearance of life-forms on Earth, an event now thought to have occurred within 100 million years of the time the planet cooled and became habitable. The evidence also suggests that life evolved rapidly after its first appearance. Perhaps the discovery that life is robust enough to adapt to environments previously thought to be sterile—in boiling water full of organic solvents and heavy metals, for example—may explain this rapid burgeoning. Or perhaps some of those life-forms were delivered by interplanetary mail.

This process need not have been restricted to our solar system. After all, how did the organic molecules on Hale-Bopp get there in the first place? One possibility is that they were cooked up in the comet itself. Hale-Bopp's large tail, extending almost 30 degrees across the sky, so far from the heat of the Sun, suggests that there may be internal energy sources in the comet itself. Perhaps the material inside its frozen shell is liquid. In such a primordial soup, might something akin to the classic Urey-Miller experiment of 1953—in which a primitive atmospheric "soup" of methane, ammonia, hydrogen, and water was zapped with an electrical current to produce various organic compounds including two of the constituents of proteins, glycine and alanine—have been carried out on a cosmic scale?

Alternatively, free-floating organic molecules may have been generated before or during the formation of our solar system. Organic molecules have been detected spectroscopically in interstellar space for some time. Perhaps the organic seeds of life are ubiquitous in the galaxy—in effect, waiting for the right conditions to settle down.

Though Mars may once have been hospitable to life, it does not appear to be now. However, at about the same time as the announcement of the putative Martian fossil life-forms, NASA released pictures taken by the *Galileo* spacecraft during a close flyby of Jupiter's moon Europa. The surface of Europa is clearly frozen, but the markings thereon indicate signficant disturbance either from internal energy sources or the gravitational tidal stress induced by Jupiter. What appear to be ice floes and evidence of geyserlike activity suggest that liquid water may well have existed, and may still exist, beneath the moon's frozen crust. And just as on comet Hale-Bopp, perhaps a slew of organic molecules exists there, too. Given the discoveries of life in unlikely places on Earth, it is even conceivable that self-reproducing life might exist in a hidden Europan ocean. Indeed, the numerous small moons of the outer planets appear to offer more potential niches for the development of life than their planets do.

Of course, as exciting as it would be to find life on Europa or on Saturn's Titan, say, it's clear that there is no intelligence outside Earth in our own solar system. If we want to find kindred spirits in the universe, we have to look beyond our Sun. While the preceding chapters suggest that the likelihood of doing so in spacecraft is remote, a number of new discoveries suggest that we may eventually be able to directly detect Class M planets—*Star Trek*'s term for Earth-like systems—orbiting other stars.

Until a few years ago, while astronomers had long argued that a significant fraction of stars probably possess solar systems, skeptics countered that though there were perhaps as many as

400 billion stars in our galaxy, there were still only 9 known planets. Well, that isn't the case anymore. We have discovered a handful of planets orbiting Sun-like stars, the closest of them tens of light-years away. The evidence suggests that planetary formation is a rather common occurrence and not at all the rarity it once appeared to be. The first claimed observation of a planet orbiting a Sun-like star other than our own was made in 1995 by Michel Mayer and his group at the Geneva Observatory. Most of the new data, however, and certainly the most convincing results obtained thus far, have been amassed by a group centered at the University of San Francisco, under the direction of Geoff Marcy, which had been carefully tooling up for this task over the past decade.

The search for planets outside our solar system has been undertaken by making use of the following idea: Although we customarily think of the Copernican Revolution as the discovery that our planet orbits the Sun rather than vice versa, this is not strictly correct. Gravity is a two-way street. The same Newtonian law that tells us that hovering giant flying saucers will crush us implies that as planets orbit the Sun, the Sun moves in response. While we tend to idealize planetary orbits by imagining them around a fixed Sun, in fact both planet and Sun orbit a point located between them, called the center of mass of the system. Because the planets are much lighter than the Sun, this point is located close to the center of the Sun, so that the Sun actually orbits a point just slightly outside its surface.

Thus (as the Catholic Church maintained steadfastly for the nearly 400 years it refused to reassess its condemnation of Galileo) the Sun orbits in the solar system! But not much. In fact, we can estimate how much by recognizing that since Jupiter is by far the most massive planet, its gravitational pull dominates the calculations. Since Jupiter orbits the Sun once every 11.86 years, this means that the Sun orbits the center of mass of the Sun-Jupiter system—situated just outside the Sun's surface, at a distance of about 800,000 kilometers from the Sun's center—once every 11.86 years. If you then work out the velocity of the Sun in this orbit, you find that it is moving about 10 meters per second, or at about the same speed reached by Olympic sprinters. For a human being, this is pretty fast; for an astrophysical object like the Sun, it is almost unimaginably slow.

A sensible person—say, a *Star Trek* writer—might dismiss the possibility of measuring motions this small in distant stars; however, one of the most fascinating things about modern experimental science, at least to me, is that precisions which once seemed fantastical are now routinely achieved. The key is not unique to planetary searches: it is the workhorse of modern astronomy—the Doppler effect. (For those whose only association with this effect is high school physics, it may lack poetry, but poetry is in the ear of the hearer. I have a cartoon in my office by the great science cartoonist Sid Harris; it shows two cowboys on the plain at sunset, looking at a distant train. One cowboy says to the other, "I love to listen to the lonesome wail of a train whistle as the magnitude of the frequency of the wave shifts due to the Doppler effect.") The well-known fact that sirens are pitched higher as they approach than they are after they've passed has been used by astronomers for most of the past century to learn about the universe. The siren sounds higher because the sound waves coming at you are of shorter wavelength, which produces a higher pitch. The same phenomenon obtains for light; when light is emitted by an object moving toward you, the waves you receive are compressed, making the light look bluish. If it's moving away from you, the light shifts to the red. The American astronomer Edwin Hubble became famous in the late 1920s for his demonstration that light frequencies emitted by distant galaxies showed that these galaxies, on average, were moving away from us, and that their velocity was proportional to their distance. In this way, we discovered that the universe was expanding.

In a similar way, by observing the frequency shift in the light from one side of a galaxy and comparing it with that on the other, astronomers can infer the galactic rotation rate. In the 1970s, Vera Rubin and her colleagues were able to show that this rotation was anomalous—that is, the galactic motions appeared to be due to the gravitational pull of a great deal more matter than was visible in the galaxies themselves. Thus was "dark matter" discovered. It turns out that over 90 percent of the mass in the observable universe is in the form of stuff that doesn't shine, and its nature is one of the outstanding puzzles of modern astrophysics and cosmology.

Clearly, the simple Doppler effect can be pretty powerful, and in 1995 and 1996, Mayer, Marcy, and their colleagues were finally able to use it to measure the wobbles of nearby stars and thus discover a new kind of invisible matter: Jupiter-size planets. In such investigations, one has to make very precise measurements not just of a star's wobbling velocity but of the period of the variations in its velocity in order to determine the characteristics of the orbiting planet. With these two measurements, the mass of the planet can be unambiguously determined.

Indeed, what is most remarkable is that some of these newly discovered giant planets, up to nearly 5 times the mass of Jupiter, seem to be in orbits closer to their host stars than Mercury is to the Sun. One of them—the first to be found—has an orbital period, or "year," of only about 4 days! Not long before these observations were made, theoretical predictions had suggested that giant planets could not form in orbits that close to their Sun because of tidal stresses. The new observations suggest that planet formation may be not only easier than previously thought but also much more varied. Perhaps our solar system is not particularly typical. With a new set of possibilities for planetary formation, new possibilities emerge for the origin of life.

It is important to stress that the planetary systems observed to date appear unable to support Earth-like, advanced life-forms. The conditions of extreme heat and very high surface gravity are unlikely to allow the evolution of such life. One of the newly discovered planets, however, is far enough from its host star so that liquid water might exist on or near its surface. As we have learned from recent discoveries on Earth, this and a little heat may be all it takes to support primitive life.

I want to emphasize how astonishing the discovery of these Jupiter-like planets really is. To infer the existence of these objects, stellar motion on the order of tens of meters per second must be observed by means of Doppler shifts. Such motions produce frequency shifts of the observed light of less than 1 part per million. Not only do such small frequency shifts have to be resolved, but they have to be carefully monitored over days, weeks, and months to demonstrate convincingly that their regularity is indicative of an orbiting planet and not, say, the ordered pulsations of the stellar surface. Because of their perseverance and technical mastery, a small group of dedicated observers has brought us one step closer to the stars.

However, as agents Scully and Mulder would probably tell you, sifting through indirect hints of alien intelligence is interesting but only enough to begin to get the blood going. Whereas coming face-to-face, or at least body part to body part, with an alien—now, *that's* what it's all about! No matter how many exotic metallic objects the *X-Files* team extracted from the nasal passages of alien abductees, it would probably take the discovery of a bona fide alien body—one that didn't keep inconveniently disappearing—to persuade their superiors (or at least the ones who aren't part of an evil government conspiracy) to pay attention. Sometimes only seeing is believing, even on *The X-Files*.

Similarly, as exciting as the discovery of extra-solar-system planets is, it's worth emphasizing that we still have not yet *seen* one directly. Moreover, the velocity kick given to a Sun-like star by an Earth-like planet at an Earth-like distance is only about 10 centimeters per second, and even indirect detection of such an object is no small task. To resolve such velocities would require a frequency resolution and stability of better than 1 part in a billion, a result unlikely to be obtained in the foreseeable future. Even if it were, so many other sources of astronomical "noise" might be picked up at this level that the signal would be hopelessly buried.

A technique that might allow us to infer the existence of smaller planets at Earth-like distances from their stars involves measuring not the velocity of motion of the star in response to the planet's orbit but the change in the star's position on the sky. This technique was developed more than 100 years ago by the first American Nobel laureate in physics, Albert A. Michelson, of the Physics Department at my home institution, Case Western Reserve University (then Case Institute of Technology). It is called optical interferometry. A distant light source is simultaneously observed by two neighboring telescopes, so that the troughs and peaks of the light's wavelength can be compared. Since the wavelength of visible light is so small, even a small change in the position of the star on the sky will produce a measurable change in how these peaks and troughs line up at the two telescopes. This allows one to obtain a high resolution of the star's motion on the sky. A new binocular device atop Mt. Palomar has a resolution on the sky in principle of about a 100-millionth of a degree. This is of an order I would have labeled science fiction just a little while ago—it's like resolving from a vantage point on Earth whether I'm holding up one finger or two while I'm standing on the Moon!

You might suspect that if we can achieve this level of resolution, we should be able to directly "see" the planets orbiting nearby stars. From there, we are just one step away from taking out our tricorders and scanning for life-forms, as Dr. McCoy or Dr. Crusher might do. Well, there's still a problem to be overcome. While in principle one can easily resolve the distance between a planet and a star if the planet is the same distance away from its star as the Earth is from the Sun, and if the system under observation is within, say, 100 light-years from us, the problem is that stars are very bright, while their planets, which merely reflect the light, are much darker. On top of this, there is a competing problem. As the light from cosmic objects passes through our atmosphere, it bends to and fro because of variations in air density, motion, and so on; as a result, the signal from a point source is spread over a disklike region. This "seeing disk" for a typical terrestrial observatory is such that the light from a nearby star would easily envelop the space containing its planets.

One of the few examples of something useful produced by work on the Strategic Defense Initiative is a technique known as adaptive optics, which has allowed astronomers to circumvent this last problem, in principle. Thankfully, now that SDI is defunct, this once-classified research is being put to good use. The idea is simple: If one has a reference object whose original light profile is known or can at least be closely approximated, then by observing this object through the atmosphere and seeing how its light is spread out, one can subtract the effects of the atmosphere at any given instant. If there is another object close on the sky to the reference object, one can use this subtraction technique to resolve the second object with a greater degree of accuracy. But what if there is no reference star close to the one you want to observe? Well, at Lawrence Livermore National Laboratory, one of the original homes of SDI research, a group has come up with a novel solution. If you don't have a star nearby, then why not just make one?

This sounds even more ambitious than something Geordi LaForge or Data would suggest to Captain Picard—or something only a research group made giddy by a surfeit of Defense Department dollars would undertake. But from an operational standpoint, a star is simply a point of light in the sky—something a whole lot easier to create than an actual star. Lawrence Livermore scientists have done it using a powerful laser based on light emitted by sodium atoms. The laser beam is powerful enough to make it up through the atmosphere as a thin column of light. About 30 kilometers above the Earth's surface, sodium atoms in the rarefied upper atmosphere absorb this laser light and reradiate it. *Voilà!*—a glowing point of light in the sky! It is amazing to see photographs of these artificial stars high above the lights of Livermore, California, at night. One can see the narrow, powerful laser beam rising into the sky; then its light fades as the atmosphere off which it reflects becomes thinner; then, high above the ground, in the region where sodium is present to absorb and reradiate the light, is a single yellowish-red "star."

Since one knows very well what the initial profile of the laser beam is and exactly where it is pointed, one can use the observed characteristics of these "guide stars" to subtract the effects of the atmosphere with great precision. And since one can shine the laser in any direction, one can place the guide star as close as one wants to the star one wants to observe. It is thus possible to model the scattered light from the real star, allowing one to probe for faint objects in its vicinity. More important, one will localize the faint light from any orbiting planet (which is also spread out by atmospheric effects) amid the smooth background of noise—the scattered light from the nearby star. As difficult as this sounds, some astronomers believe that within a decade—if the Keck 10-meter telescopes in Hawaii, the largest telescopes in the world, are fitted with a laser guide-star apparatus—it will be possible to directly observe the dim light of Jupiter-like planets. One of my colleagues at Case Western Reserve University, Glenn Starkman, has added a new wrinkle to this scheme. He proposes sending up a satellite that will release a large balloon, which can then be maneuvered to occult the ambient starlight and thus aid in the planetary search.

Once the prospect of directly observing planets becomes possible, the idea of scanning for life does not seem all that far-fetched. Of course, one would not look for life-forms directly; however, by observing the color of light reflected by a planet, one can learn a great deal about its atmosphere and the characteristics of its surface. NASA has proposed direct observation of extra-solar-system planets as one of the agency's goals for the next century. The next generation of telescopes in space will build on the incredible success of the Hubble Space Telescope, surpassing any observations we can now make from Earth, and I am prepared to believe that within the next century we may well directly detect the existence of an organic, water-filled world elsewhere in the cosmos.

CHAPTER SEVEN

Gambling on the Galaxy

He is glorified not in one, but in countless suns; not in a single earth, a single world, but in a thousand thousand, I say in an infinity of worlds.

—*Giordano Bruno*

Thankfully, only 400 years after Bruno was burned at the stake for this claim, screenwriters are relatively free to let their imaginations run wild. I have always been impressed with the ingenuity of Hollywood science fiction writers when it comes to the creation of alien beings. Yet if there is one place where literary and back-lot imagination probably fall short of the mark, it is in conjuring up the possible variety and quantity of life in the universe. Even when you put together the silicon-based Horta, the insectlike Harada, and the cyber-based Borg; the Wooky, the Sand Worms, Yoda, and Jabba the Hut; little ET, and the slimy beasts of *Independence Day* and the *Alien* series; and all the creatures in *Men in Black*, you barely scratch the surface of what may be possible.

Consider the following. In DNA-based self-replicating life-forms, there are 4 different genetic "letters," and approximately 1,000 of these letters, in various combinations, make up a gene; you therefore end up with approximately 10^{600} possible variants. Even if nature somehow produced a new gene combination once every second in each cell on Earth throughout Earth's history, the total number of combinations thus produced would have been only about 10^{47}.

Now, many of the individual letters in a gene may be irrelevant, but even so, if 99.9999999999999999999999999999999999-99999999999999 percent of all the possible gene combinations lead to junk genes, the total number of different life-forms which could have appeared on Earth this way would still be smaller in relation to the number of viable possibilities than one atom is compared to the total number of atoms in the universe!

And that's just DNA. We have no idea whether other self-replicating organic, or inorganic, combinations might also be able to exist—in which case, the above estimate of the possible varieties of life in the universe could be too small by many orders of magnitude. Not only are the possibilities virtually endless, but a host of exciting discoveries in recent years have caused us to readjust upward our estimates of how likely life might be to evolve elsewhere in our galaxy. If there is to be a Year of the Alien, this past year is one of the best candidates so far. Every indirect indication we have suggests, now more strongly than ever, that life is ubiquitous. We once had no notion at all of how the building blocks of life might have formed on Earth; now we have a variety of compelling competing theories. Moreover, as I noted in the last chapter, life has been discovered in all the wrong places. Nothing is more exciting for a scientist than when things turn out as we didn't expect, amid a wealth of new data.

It is worth stressing that not everything is possible. In spite of the great potential diversity of possible extraterrestrial life, the writers of *Star Trek, The X-Files,* and the *Alien* movies (and some putative UFO abductees and their psychiatrists) have alas sometimes overshot the mark. An example is the screenwriterly propensity for portraying successful interspecies mating. (By this I don't mean interspecies coupling, which happens every now and then on Earth, and with reckless abandon on *Star Trek*. There is the famous scene that escaped the censor's notice in the 1960s between Captain Kirk and Queen Deela; there's the love affair of Dr. Beverly Crusher and Ambassador Odan; and the dalliances carried on between the virile Commander William Riker and almost every alien in a skirt.) Of all the nonphysics issues about which I received letters after my last book, this aspect of *Star Trek* seems to have provoked the most scorn—although I suspect that the human-alien hybrids on *The X-Files* generate less wrath among its viewers. On Earth it is well known among biologists and some farmers that copulation between species rarely produces viable offspring. The genetic code, while apparently infinitely malleable, is also quite sensitive. You might as well try running a Macintosh code on a Windows 95 system! Even species that are remarkably close in genetic makeup are biologically incompatible in matters of reproduction. And in the rare cases where offspring are viable—mules, for example—they themselves generally cannot reproduce.

Now, this is true of species that have coexisted on the same planet for perhaps millions of years, and have responded to similar sets of evolutionary imperatives, with genomes that are not markedly dissimilar. Imagine attempts at cross-breeding between two species that have evolved on separate planets. Even if the fundamental chemistry was the same—something not necessarily likely—it's extremely difficult to believe that the product of mating, say, a Vulcan and a human being would produce anything as viable as Mr. Spock, any more than the coupling of a human and a chimpanzee would be likely to produce successful offspring. (My ordering should not be taken to suggest a correspondence to the Vulcan-human analogy.)

In any case, the fascinating new discoveries of the past few years have changed the way we think about the probabilities of life in the cosmos. Previously, the existence of planets outside our own solar system was pure speculation, and the range of conditions that might allow life to form and survive was thought to be far narrower than we now know it to be. At no time in this century has there been more reason for optimism about the possibilities of discovering extraterrestrial life, perhaps even intelligent extraterrestrial life, in our future.

For over 30 years, the standard estimate for the probability of the existence of extraterrestrial civilizations has been codified in what became known as the Drake equation—after the astronomer Frank Drake, who proposed it. In this equation, the number of intelligent civilizations in the galaxy is calculated as the product of the number of stars in the galaxy times several different probabilities expressed as fractions: the fraction of stars that are likely to have planetary systems; the fraction of these that are likely to have Earth-like planets; the fraction of such stars that are likely to be stable long enough for life to evolve; the fraction of these life-forms that are likely to evolve to achieve intelligence, and so on. In a sense, what this equation does is parameterize our ignorance, since each of the fundamental probabilities that goes into it is subject to debate. In this way, different groups have estimated the number of intelligent civilizations in our galaxy as ranging from millions to one. However, as time goes by and our knowledge increases, more reliable estimates for at least some of these factors have emerged.

Nevertheless, I have always felt that there is an inherent problem with this approach, and I recently had a discussion about it with Frank Drake himself, at the Naples conference on the search for extraterrestrial intelligence, which I mentioned in chapter 2. The point is that many of the individual probabilities whose product goes into the equation are small, and their product is even smaller. Thus, one goes from perhaps as many as 400 billion stars in our galaxy to perhaps only a handful of intelligent civilizations. Now, when probabilities get this small, they are sometimes difficult to estimate. The statistics of very rare events is quite subtle, and the most naive application of probabilities may not be the best way to approach this subject.

In the first place, whenever one considers a probability that results from the product of many different individual probabilities, the result has to be a small number, because each individual probability that goes into the product is less than 1, and the product of many numbers smaller than 1 is always very small. For example, the probability of any one particular event in your life taking place is, when viewed this way, almost zero. The probability that I woke up this morning in Geneva at 7:30 A.M. required first that I be on leave from my home institution, which in turn required that I be at that institution in the first place, which required that I chose physics as a career, and so on. More immediately, my waking up at 7:30 A.M. probably required that there be a small pond outside my window, in which a particular tadpole had become a frog that croaked at 7:29 A.M., and so forth. Though all these probabilities (and others too numerous to mention) were small, leading to an infinitesimally small probability that I would do exactly what I actually did, nevertheless I actually did it. Events with small probability happen all the time, because *all* events, when viewed in this way, have small probability.

By the way, this is one reason we have to be careful when someone tells us something like the following: "I had a dream the other night that my wife cried out to me as she fell down the stairs and broke her leg. A week later, she did trip and injure herself—isn't that amazing? The probability that my dream would come true is so small that something fishy must have been at work here." Well, to this notion the famous physicist Richard Feynman used to have an interesting rebuttal. He would sometimes exclaim, "You'll never believe what happened to me this morning!" When you took the bait, he would answer, "Absolutely nothing special!" The point is that we tend to remember those events that stand out and forget those that don't. An amazing coincidence is in any event amazing, but perhaps not as amazing as we might think.

There is a related problem one must confront here. If one considers the probability of many separate events occurring, one must also consider whether or not they are correlated—that is, whether or not they are truly independent. If they are correlated, simply multiplying individual probabilities will not give you the correct estimate, and the final probability may actually be much larger than one will predict if one makes this error. For example, the probability that I will utter an obscenity at any given instant may be small (although it is certainly not zero). The probability that I will hit my funny bone at any given instant is also small. However, the probability that I will hit my funny bone and then utter an obscenity is not equal to the product of the probabilities, since the probability of swearing at a given instant is correlated to the probability of hurting myself at a given instant. Similarly, the probability that a planet might survive meteoric and cometary impacts long enough for intelligent life to evolve may be small. And the probability that a solar system has a Jupiter-size planet in its outer reaches may also be small. But these two factors are not independent: The gravitational effect of Jupiter is believed to be important in deflecting many potentially lethal objects away from Earth's orbit.

The modern parlance for these notions is "conditional probability." Its expositors hold that we should not concern ourselves with "absolute probabilities," which often have no relevance to things as they are, but with "conditional probabilities"—the chances that some event will occur when some set of previous conditions exists. However, we don't always know what probabilities are conditional on other probabilities; as a result, things can get complicated when you try to estimate an exact probability that some specific complex set of events will occur—outside of those performed in controlled experiments in the laboratory.

A way has been developed around this problem, based on the somewhat nonintuitive but extremely important notion that something that has a small absolute probability can nevertheless happen more frequently than any of the possible alternatives. As I've mentioned, every event that happens in the world can be viewed as having a vanishingly small probability if all the contingent factors are taken into account. What is therefore important is not the absolute probability but the relative probability. Given a wide variety of outcomes, what set of observations is more likely than others? If one set of possible outcomes has a raw probability of 1 in a million—well, that sounds pretty small. But if the other millions of sets of outcomes each have a probability closer to 1 in a billion, then the first set of outcomes is 1,000 times more likely to be observed in a single trial than is any other set of outcomes.

Of course, where so many possible outcomes are involved, what becomes operationally important is not one specific set of outcomes so much as whether the observed set is close to the one with maximum likelihood. An example should make this clearer. Say that I begin a series of coin flips and count the number of heads and tails. We all intuitively know that the maximum likelihood is that the number of heads will be approximately equal to the number of tails. However, we don't expect the number of heads *always* to equal the number of tails. If I flip the coin 10 times, I may get 6 heads and 4 tails, or vice versa. As I flip the coin a larger and larger number of times, the number of different sets of possible outcomes continues to increase, and thus the probability of any specific set of outcomes (say, 499 heads and 501 tails out of 1,000 flips) gets smaller and smaller, precisely because there are more and more different possibilities that can occur. But despite the fact that the absolute probability of any specific combination decreases, the relative probability of getting very close to 50 percent heads and 50 percent tails gets higher and higher. By the time you have flipped a million times, the likelihood of deviating from this mean value by only 10 percent is 1,000 times smaller than the likelihood of lying within 1 percent of a 50–50 split! This is true despite the fact that the probability of getting 500,000 heads and 500,000 tails, the most probable outcome, is less than 1 in 1,000.

I can write down any specific tally of heads and tails that may result when I flip the coin a million times. It's easy to calculate the probability of this specific set (say, HHTHTTTHTHT . . .) occurring, since there's only one way it can occur. Since with each flip the probability of a head, say, is .5 (that is, 50 percent), the probability of getting the sequence in question is $(.5) \times (.5) \times (.5) \ldots = (.5)^{1,000,000}$, which is, needless to say, a very small number.

So, since each specific sequence—even a T repeated a million times—has precisely the same probability as any other specific sequence, how come we never expect that at the end of the million flips we will have a million tails? Well, because there are many different ways of writing down a sequence of Hs and Ts that will end up with 500,000 Hs and 500,000 Ts, but there is one, and only one, way of writing down a sequence of a million Ts. It's as simple as that.

What the technique of maximum likelihood does in this case is to find the characteristics of those types of sequences which have the maximum likelihood of occurring by comparing relative probabilities, without worrying about absolute probabilities, and also recognizing that any one particular sequence may be extremely rare. The method in this case would tell us that the sequence resulting in something close to 500,000 heads is much more probable than anything else, so that one of the possible sequences leading to this result is more likely to be observed than anything else, even though the probability of any one particular sequence occurring is extremely small.

Now, what is the point of all this when it comes to the possibility of extraterrestrial life in the universe? Well, what may be important to consider is not the absolute probability of any specific sequence of events leading to intelligent life, but rather the relative probability of some such sequence occurring compared to the probability that some sequence will occur which will *not* lead to life. It is the relative probability that is important. If we have learned anything over the past decade, it's that life is more robust than we had imagined. I'm now more willing to assume that when you have organic material in the presence of some heat, some light, and some water, it's difficult for life *not* to arise, even if the probability of its arising by any specific sequence of events is small. Instead of considering how probable it is that Earth-like conditions would obtain on any other planet, it might be more appropriate to ask, "What is the probability that organic materials will *not* by *any* route form self-replicating systems in several billion years on a given planet?"

I repeat that I have no idea of the answer to this question, and I emphasize that the answer lies outside my expertise. But it seems to me that, as in the coin example above, there could be many more routes to the evolution of life-forms than there may be to ensuring that a given solar system is devoid of life.

Once one thinks in these terms, focusing on the remarkably lucky specific series of circumstances that led to the evolution of intelligent life on Earth may be wide of the mark. If the likelihood of some type of life evolving on some system is greater than the chances of ensuring that no life at all arises, then—as remote as the probability of any particular sequence of events leading to life may be—we are more or less guaranteed that some such sequence will occur in most situations.

I am not suggesting that the Drake equation is flawed as it stands—it is not—nor that it needs on fundamental grounds to be replaced by the Krauss equation, even if that does have a nice ring to it. If we knew all the contingent factors leading to any type of life, we would be able to write down the probabilities exactly, and thus accurately determine the number of intelligent civilizations. And maybe one day we will be able to, since evolutionary biology is itself evolving by leaps and bounds. In advance of that knowledge, though, comparing relative probabilities may provide us with better insights.

Finally, there is an overriding factor suggesting that the formation of life—even intelligent life—may be possible or even common elsewhere. It is that we exist. This undoubted fact demonstrates that intelligent life can form under at least some subset of circumstances known to be present in the galaxy. Moreover, the lessons of natural history on Earth suggest that not only is life extremely robust, persisting even through mass extinctions, but also that the evolutionary routes leading to different complex organisms are numerous. In this regard, one should note with caution that while natural history tells us that life formed relatively quickly on Earth, it still took almost 4 billion years for *intelligent* life to evolve, and even then only by a series of historical accidents. This could well mean that life is common but intelligence isn't. On the other hand, by the same argument given above, intelligent life might result from many different historical trajectories, and the one that produced us might be only one of many. Hard to know with a sample of only one!

In general, I suspect that since our own Sun is a rather ordinary star, and its place in the galaxy is unremarkable, and since nature repeats herself as often as the laws of physics and chemistry allow, it would be odd if life weren't ubiquitous in the galaxy. It is just a matter of time—although perhaps on a cosmic timescale—not a matter of principle, I believe, before we discover our galactic cousins. I'll go even further and say that I expect microscopic forms of life to be found elsewhere in our solar system within the next century. (Whether they will turn out to have a common origin with life on Earth is an open question.) The discovery of extraterrestrial intelligence, however, is doubtless much farther in the future, simply because of the near impossibility of round-trip travel to the stars, and also the difficulty of communicating across the vast abyss of space in the absence of agreed-upon forms of communication.

Look at it this way. Even without boats to travel across the Atlantic or Pacific, it is possible to send messages, or at least greetings, to civilizations on the other side of the world. Messages in a bottle have been discovered, for example, thousands of miles from their origin. Yet for about as long as it took European civilization to evolve to the point of transatlantic travel, there was no knowledge whatsoever of New World civilizations.

But unlike the first transatlantic explorers, who set sail with the intention of bringing back riches to their homelands, Earth's first interstellar travelers will probably have no intention of returning. Like many a refugee, we will move out into the galaxy because we will have no other choice. The laws of physics, not the laws of mankind, will require us to leave.

CHAPTER EIGHT

The Restaurant at the End of the Universe

The choice is: the universe . . . or nothing.

—*H. G. Wells*

With the approach of the millennium, now is an appropriate time to join the crowd and proclaim the inevitable: The world *is* going to end. When—and, perhaps more important, how—are issues that are not quite so clear.

On the whole, I think Doomsday has gotten bad press. I will argue here that it holds great potential for the human race. With typical astronomical precision, we can pinpoint an upper limit for human existence on Earth at about 5 billion years from now, give or take 500 million years. So there's still time to get your broker on the phone and sell your stocks, and don't give up that reservation in Paris for Christmas 1999. Still, in the language of logicians, this upper limit, which marks the whole planet's termination, is a sufficient but not necessary date for our demise. We could easily perish long before, in a global Armageddon, or because of some new efficient virus, or because of an astronomical catastrophe such as a large meteor impact. Or, of course, aliens could decide to annihilate us.

When the aliens of *Independence Day* started their rampage, they had a clear goal—to exploit the Earth's resources, before discarding it like an old apple core. Compare this with one of my favorite doomsday devices (along with Stanley Kubrick's whimsical creation in *Dr. Strangelove)*—the neutronium planet killer in the classic *Star Trek* episode "The Doomsday Machine." This machine destroyed the civilization that created it, and once there was nothing at home left to destroy, it wandered through the galaxy, offing whatever other planets it found. I am particularly taken with this idea because of the complete purposelessness of the destruction; I fully expect that this will be the nature of our own planet's end.

Here's a tricky question. If I were to turn off the nuclear reactions that power the Sun (as, for example, was done to an unfortunate star in *Star Trek VII: Generations*), how long would it take for the Sun to stop shining? The right answer is surprising. In the nineteenth century, two giants of theoretical physics, Lord Kelvin in England and Heinrich Helmholtz in Germany, each tried to determine the answer. Their question was equivalent but slightly different, because no one at the time knew anything about the nuclear reactions that power the Sun. Kelvin and Helmholtz both assumed that the source of the Sun's heat was its own ongoing gravitational collapse, and that it was gradually shrinking and cooling as it radiated heat. If the Sun's own mass was its power source, they wanted to know how long the Sun would burn after it was first formed. The answer they derived was between 30 million and 100 million years.

This was a truly amazing result, I think. It implies that if I were to turn off the Sun today, it would continue to burn for at least 30 million years, powered by gravitational collapse alone, before dying out like an ember! (The sudden shutdown in *Generations* would never have happened, but the reality would have been far too slow to interest even an audience of dedicated Trekkers.) Thirty million years or so may seem like an awfully long time, but in fact it got Kelvin and Helmholtz into hot water (forgive the pun). Since they didn't know of any internal power source, they reasoned that because the Sun was still shining, it definitely had to be less than 100 million years old. The problem with this was that even at the time Kelvin and Helmholtz did their calculations it was known from fossil evidence that the Earth was much older than 100 million years.

Creationists love it when sound scientific reasoning produces a cosmic paradox. But what such paradoxes really provide is an opportunity for discovery. The fact that the Sun had to be at least as old as the Earth suggested that there was an internal power mechanism that kept it shining. Fewer than 50 years after the Kelvin-Helmholtz estimate, natural radioactivity was discovered in the laboratory, and fewer than 50 years after *that,* nuclear power was harnessed. In 1938, the great theoretical physicist Hans Bethe, who is still alive and calculating today, finally showed that nuclear reactions could power the Sun, a theoretical discovery for which he later received the Nobel Prize.

Incidentally, for those of you who like to debate with the creationists who believe that the solar system is only between 5,000 and 7,000 years old, here is some useful ammunition. Guess how long it takes radiation emitted deep inside the solar core to make its way to the solar surface? Again, the answer is surprising: it takes almost 10,000 years! The reason is simple. The Sun is very big: its radius measures some 432,000 miles, and a quantum of radiation emitted in its interior travels on average about 1 centimeter before it hits something and is scattered in another direction. The random walk that ensues takes about 10,000 years (again, on average) to progress to the surface. This means that if the Sun were only 5,000 years old, it would not yet be shining—

at least not with anything like its present consistent brightness!

It is these two factors—the random walk and the competition between gravitational contraction and nuclear burning—that determine how fast the Sun will burn its nuclear fuel. And it is this rate of nuclear burning that sets a limit on the term of life on Earth.

Since the origin of the solar system some 4.5 billion years ago, the Earth has been a slave to the Sun. Every process, every major event in our terrestrial history, has been dependent on our closest cosmic stellar companion. The average energy received on Earth every day from the Sun is tremendous—about 1,350 watts per square meter. Day in, day out, for 4.5 billion years the Sun has been bathing the Earth with almost 100 million billion watts of radiation. This solar radiation makes life possible on Earth, but it takes its toll on the Sun.

The 100 million billion billion total watts of power the Sun has been pumping out since its formation is, as we have seen, not directly due to the energy released by collapsing dust and gas. Instead, relentlessly (and, when viewed on an atom-by-atom basis, slowly), more than 10^{38} hydrogen nuclei in the solar core are converted each second into nuclei of the next lightest element, helium—enough nuclear reactions to power almost a million 10-megaton hydrogen bombs per second. The incredible pressure generated by these reactions is enough to balance the gravitational attraction that would otherwise cause the Sun to collapse inward.

As a result of this nuclear burning, the solar core is inexorably converting from mostly hydrogen to mostly helium. As the relative abundances change, the whole structure of the Sun changes in response. Over the course of a human lifetime, this change is not noticeable (although there are some changes that are, like the sunspot cycle, whose 13-year periodicity is still not understood). However, over cosmic time the Sun's structure has changed considerably. Since life arose on Earth, the luminosity of the Sun has increased by almost 25 percent, for example. And so too, eventually, the Sun will run out of its hydrogen fuel. In spite of the complexity of the reactions taking place in the solar interior, it is a relatively straightforward matter to determine when the hydrogen fuel will run out: simply divide the total energy the Sun produces per second by the energy produced each time 4 hydrogen nuclei fuse to form a nucleus of helium, in order to get the number of hydrogen atoms being burned per second, and then divide that by the amount of hydrogen left in the Sun's core. The answer is about 5 billion years.

Unlike many larger stars, however, the Sun will not end its life with a bang but with a whimper. The exhaustion of its hydrogen fuel does not leave the Sun with nothing to burn. Helium is itself susceptible to nuclear burning, at a much higher temperature, to form yet heavier elements, such as boron, carbon, oxygen, and nitrogen. How does the core of the Sun—the region in which nuclear burning takes place—reach the higher temperatures at which the nuclear burning of helium can take place? Simple. The core contracts because of the Sun's own gravity, and the pressure and temperature of the gas inside increases in response until the temperature for helium burning is reached.

Now, the rest of the Sun does not stand idly by while all this excitement occurs in its core. During the final stages of hydrogen burning, as the core begins to contract, the Sun's outer layers puff up, due to the extra release of heat from the core. The size of the Sun will increase many times, turning it into what is called a Red Giant. While this is merely one of a series of metamorphoses the Sun will experience during its lifetime, it is a particularly important one for Earth, since in the puffing-up process the solar surface will increase enough to envelop the Earth's orbit. From then on, our tiny speck in the solar system will be no more.

It may sound fantastic to think of the Sun puffing up by such a great factor, so let me tell you something even more fantastic.

The largest known star is Mu Cephei, which has a radius of 11 astronomical units. An astronomical unit is the distance from the Earth to the Sun—93 million miles—so this star would encompass our solar system out to Saturn. I find that remarkable to contemplate.

The Sun will continue to evolve after it gobbles the Earth; eventually, helium will burn to form carbon, carbon will burn to form oxygen, and so on, until the nuclear fusion process reaches iron. At this point, nuclear burning stops, since iron, the most tightly bound nucleus in existence, cannot release energy by fusing to form a heavier nucleus. Thus the nuclear fight against the force of gravity will be lost, and the Sun will collapse inward to form a dark star known as a white dwarf, slowly radiating away its stored energy. Eventually it will die out like an ember and join the blackness of space around it. What remains of the Earth will have merged with its stellar host; like a member of the Borg collective, Earth will have utterly lost its identity.

This ultimate calamity is so far removed that it's pretty well irrelevant to us, our children, our children's children, and their children . . . and on down the line. But even if we are lucky enough to survive all the other challenges to our continued existence, our species' days are numbered.

That is, of course, if we remain on Earth—a Big IF. I imagine that we will have chosen to leave well before the Sun blows up, if our species persists long enough to develop the means to leave—another Big IF. Since the burgeoning speciation at the dawn of the Cambrian, some 540 million years ago, there have been 5 mass extinctions, during which a significant fraction of species alive at the time disappeared. The largest occurred at the end of the Permian, around 250 million years ago, when up to 96 percent of all species on the planet became extinct. The most famous extinction is surely the one in which the dinosaurs perished, 65 million years ago, at the boundary between the Cretaceous and

Tertiary periods. There is good evidence that this extinction followed a collision between the Earth and an extraterrestrial object, probably an asteroid or a comet.

In order to explain these extinctions, biologists, geologists, and physicists have been examining all possible causes, and more candidates are being discovered all the time. The list of plausible potential threats to life on Earth is getting long enough so that one wonders how we have managed to survive thus far. How might we go? Let me count the ways:

1. Human Folly: This is the most immediate threat, although it may not be a global one. By this I mean that even in the event of a global thermonuclear exchange, some humans (and many other species of life) may survive. The conditions under which the unfortunate survivors will eke out their existence will be ugly, but such is life. A more deadly threat, I believe, is posed not by global war but by global complacence. We are currently polluting our water, filling our atmosphere with greenhouse gases, reproducing our numbers without regard to Earth's resources, and so on. The changes we are making act slowly on the scale of a human generation, but when you add it all up, we are in the middle of the biggest mass extinction in the Earth's history; close to 30,000 species a year are becoming extinct. We appear to be doing a better job of this than any of the natural disasters that have occurred since the Cambrian. We are unlikely to entirely wipe out our own species by this global complacence, but we may make life on Earth so unpleasant that it is preferable to leave.

2. Extraterrestrial Impact: As noted earlier, the collision of a large asteroid or comet with the Earth is the current best candidate for the Cretaceous-Tertiary mass extinction. While comparable impacts are rare, with a frequency of perhaps one every 100 million years, they are also inevitable. The advance notice we might receive of the approach of such an object would probably be months or years; we might by then possess the technology to destroy the intruder before the collision. If we don't, the Earth could become virtually uninhabitable for humans.

3. Supernovae: When a star 10 times as massive as our Sun reaches the final stage of nuclear burning to form an iron core, the gravitational pressure becomes so great near the center that the interior of the star collapses in mere seconds to form an incredibly dense object known as a neutron star. In the process, the outer shell of the star is blown off in one of the most spectacular fireworks demonstrations in the universe. The brightness of a supernova can exceed, for a period of days, the brightness of an entire galaxy. There are thought to be two or three such explosions in the Milky Way galaxy per century. The reason we don't usually see them is that (surprisingly) in spite of their intrinsic brightness, the dust in our galaxy obscures the visual signal. The last supernova to be recorded in our own galaxy was observed in 1604 by the great Johannes Kepler. Now, our Sun travels around the outskirts of our galaxy at about 200 kilometers per second—fast enough to perform a full revolution every 200 million years or so. During this time, our stellar neighbors change. If, in the course of our trip around the center of the galaxy, we were to pass within even a few tens of light-years of an exploding star, the results could be traumatic, to say the least: Earth might be knocked out of its orbit—or vaporized. Advance warning of an impending nearby supernova might be possible, depending upon the observing technology of our civilization at the time—although it is difficult to imagine what might be done to protect us from the consequences, if we are around to experience them.

4. Neutron Star Collision: Neutron stars formed in supernovae are sometimes found orbiting in binary systems, either with another neutron star or a star that is still burning its nuclear fuel. Sometimes—perhaps once every 100,000 to 1,000,000 years in a given galaxy—these two partners, losing energy and spiraling inward, will collide in a massive fireball. This may sound so infrequent as not to matter. However, over the past couple of decades, satellites originally designed for the Cold War monitoring of possible nuclear weapons tests have scanned the skies for X rays and gamma rays (which are more energetic than X rays). The results have been surprising. Short gamma-ray bursts of very high energy, lasting from seconds to days, have been observed all over the sky. Because of their uniform distribution, astronomers speculated that they were at cosmological distances—that is, not confined to our own galaxy. In this case, the energy they release would be tremendous. The hypothesis that they are cosmological was confirmed in 1997, when Caltech astronomers observed a visual counterpart to a gamma-ray burst during the burst phase and determined that this object was some 2 billion light-years from us. The best current explanation for the phenomenon involves collapsing neutron-star binary systems. It seems lately that each time a new class of energetic astrophysical object is discovered, someone speculates on a possible link to mass extinctions on Earth. One such group has calculated that if a neutron-star binary system collapsed in our region of the galaxy (perhaps once every few 100 million years), the high-energy cosmic rays released by the event would provide a lethal radiation dose to most of humanity.

5. Old Age: Finally, barring any extreme catastrophes of the type mentioned above, the Earth may become inhospitable to life simply by evolving in its own quiet way. For example, its molten iron core is thought to be responsible for the magnetic field that surrounds the planet. This magnetic field deflects most potentially harmful cosmic rays. As the Earth cools, its core will cool with it. Once the core solidifies, the charged currents that now flow to create the magnetic field will disappear. Whether this will take longer than the 5 billion years we have left on Earth is not yet clear.

And let's not forget our friend the Moon. As I noted in chapter 1, its tidal forces are ever so gradually slowing the Earth's period of rotation. Over the course of billions of years, the length of Earth's day will increase until it coincides with the Moon's orbital period. Earth's climate is bound to become unlivable long before this synchrony is reached.

Well, if you gotta go, you gotta go. If humanity is to survive these disasters, we will have to embark on a cosmic voyage to another world, or build our own and travel around in it. As I hope I've made clear, almost all the barriers to interstellar travel discussed earlier were based on round-trip travel, on the timescale of a few generations. Once we decide to leave Earth forever, the requirements change considerably. Speed is not an issue, for example. If we are heading nowhere in particular, then it doesn't matter how fast we get there. What we will require is a self-sustaining environment large enough to generate artificial gravity by rotation and to shield us from harmful cosmic rays (or powerful enough to generate large magnetic fields to deflect them). These are no small requirements, but I like to think that with several million years to get ready, even creatures with as notoriously little foresight as human beings might be up to the task.

Which brings me back to *Independence Day* once again. Perhaps 15-mile-wide ships are impractical to zip around the atmosphere in. But just as planet Earth is a self-sustaining spacecraft as it travels around the Sun, which in turn travels around our own galaxy every 200 million years, the man-made spaceships of our future, in which we will venture beyond our solar system, may also be mammoth systems, designed to house not one generation but thousands, and designed not for combat but for survival. I trust that if our spaceships make it to a safe harbor across far reaches of the cosmic ocean, we will present ourselves more generously than the visitors in *Independence Day*.

However, it is equally likely—or perhaps more likely—that the resources and the organizational and logistic skills necessary for us to leave our world in one piece will remain beyond our grasp. Will all remnants of our existence then perish with us, and with our Sun? Not necessarily. A large comet, or astrophysical shock wave, striking the Earth would not only do incalculable damage but would also eject a great deal of matter into space. Among this stuff would undoubtedly be the organic materials that provide the blueprint of our existence. Just as the organic basis of our DNA may derive from interstellar pollution, perhaps one day we will bequeath our own organic material to the universe.

One of the most remarkable astrophysical facts I know of is that essentially every atom inside our bodies was once inside an exploding star. The carbon that permeates our bodies, the oxygen and nitrogen we breathe, were not around when matter first formed. These elements were created in the nuclear furnaces of stars. In order for us to exist, it was necessary for generations of massive stars to live and die. During the fiery supernovae that marked the death of such objects, all the heavy elements that make up everything we see around us were spewed out into the cosmic nothingness. Eventually, some of this material merged with the collapsing cloud of hydrogen gas and dust that would form our own solar system. Some sort of life-forms may well have been sacrificed in these explosions, providing a part of the necessary raw materials, so that we might one day evolve and flourish. Perhaps, in one way or another, we may someday return the favor.

SECTION TWO

Madonna's Universe

We are living in a material world, and I am a material girl.

—*Madonna*

CHAPTER NINE

May the Force Be with You

> You gotta love this place. Every day is like Halloween!
>
> —*Fox Mulder*

Early in the first film of the *Star Wars* trilogy, Obi Wan Kenobi urges Luke Skywalker to "feel the Force!" To no one's surprise, Luke does, eventually, and it is very very good to him. It was also very very good to George Lucas. A billion dollars and 20 years later, the Force is still with us.

Tell me that you have not, at some time in your life, looked up at the night sky and shuddered at the vast loneliness of our existence. Or sitting alone in a darkening room, perhaps in a remote cabin in the woods, have you never, as a barely perceptible chill breeze brushed your skin, had an idea that there might be some "thing" in the room with you, which you cannot see? What *are* the things that go bump in the night?

Dark side or not, there's something particularly cozy about an invisible Force that ties the universe together and gives it meaning, coherence, legitimacy. Pondering the existence of aliens may be how we ease our innate human loneliness nowadays, but pondering the existence of invisible forces is nothing new. Such musings are, after all, at the heart of most of the world's religions, whose annual gross stretches back for millennia and makes Lucas's look like chicken feed.

In fact, invisible forces are not merely the stuff of revelation: they *are* everywhere! Turn on your radio, and suddenly there is music, borne by invisible radio waves. Leap into the air, and the force of gravity pulls you back to Earth. Pluck a couple of magnets off the refrigerator and feel them push away from each other. As a matter of fact, there is almost no such thing as a *visible* force! I say "almost" because, of course, if a piano falls on your head, the source of the force you feel (before you feel nothing anymore) is eminently visible! Or is it? What is it about the piano that makes it "material"? Why does it crush your skull?

This might seem like a silly question; after all, what could be more solid than wood, ivory, metal, all the things from which a piano is fabricated? Well, a piano, at the fundamental level, is made of billions and billions of atoms. You can therefore reasonably assume that the particles in the atoms in the piano smack up against the atoms in your head and the multiple collisions are what cause one of these atomic aggregates to spatter.

Ah, nothing could be farther from the truth. No particle in any atom in the piano—no proton, neutron, or even electron—ever gets close, on an atomic scale, to any particle in any atom in your skull. Most of what we like to think of as "matter" is actually empty space. The region in which electrons orbit an atomic nucleus is more than 10,000 times as large as the nucleus itself. It's the invisible electric forces emanating from the charged particles in the atoms in the piano that repel the charged particles in the atoms in your head and do such a good job of making both your head and the piano seem solid.

The physicist Richard Feynman used this idea to relate the

strength of the electric force to the gravitational force. I will repeat his argument here, changing it slightly so we can continue to speak in terms of your head and the piano. But instead of dropping a piano on your head, let's drop your head on a piano from, say, 100 floors up. Let's assume you are at the top of the Empire State Building, which I seem to remember from my youth has 102 stories. And say that you manage to climb over the high fence around the observation deck and do a swan dive toward the ground below. At the same instant, some piano movers have taken a union-required break from their chore of moving a new concert grand into the lobby of the building. The piano is still in several pieces, which are lying on mats on the sidewalk. Suddenly the movers look up, and to their horror they see you hurtling earthward. You land on the the instrument's elegant, polished wooden lid, which is lying flat on the ground.

Now, says Feynman, gravity has been accelerating you for 102 stories, but you don't continue your descent toward the center of the Earth: The electrical force—in this case between the atoms in the lid (in turn supported firmly by the sidewalk) and the atoms in your head—stops you cold in a fraction of an inch! Despite its spectacularly noticeable effects, gravity is the weakest force in nature.

Even this example doesn't do justice to how weak gravity really is compared to the electric force. Here's another one: Take a single electron, which has a small electric charge associated with it. If I put another electron near it, they are repelled by the electric force between them. In empty space, where no other forces were around to balance this force, they would fly apart. Now, say I wanted to pin the second electron down by putting a large mass on top of it, so that the gravitational attraction of the large mass (plus the electron) toward the original electron would exactly balance out the electric repulsion between the two electrons. How big a mass would I need?

When I asked my wife this question, she asked how far apart

the two electrons were, which is a good question. However, in this case it is irrelevant, because both the electric force and the gravitational force vary the same way with distance, so if they balance out at one distance, they will balance out at all distances.

In any case, the answer is nothing short of flabbergasting. Plugging in the relative strengths of gravity and the electric force, it turns out that the mass you have to put on top of the second electron to counteract the electric repulsion is—get this!—5 billion tons. This is not only more massive than either the Empire State Building or the twin towers of the World Trade Center, or any other Manhattan skyscraper, it is more massive than all of them put together!

Even though I have been for some time familiar with the relative strengths of gravity and the electric force, I was surprised by this particular result after obtaining it—so much so that I had to check my calculations three times and then ask a graduate student who happened to be walking by my office to check them to make sure I hadn't done something foolish. This time, I hadn't.

Why, you may naturally ask, don't we just use small electric charges to levitate buildings or large flying saucers? The answer is that these objects, if they are at the Earth's surface, are not merely attracted downward by the gravitational force of the single electron that one might hope to levitate them with, they are attracted by the whole Earth. And since the Earth is massive indeed, their "weight" at the Earth's surface is enormous compared to the force of electrostatic repulsion between electrons located any reasonable distance apart. On Earth, all these skyscrapers are extremely heavy, but in empty space they are nearly weightless. The reason that all of these skyscrapers combined are needed to balance the electric force with gravity in empty space is not that this electric repulsion is so great but that the gravitational attraction of the electron on each of these objects is so small.

Gravity is so weak that it is almost miraculous that we can detect it at all. The reason we "feel" gravity is that although the pull of each individual atom in the Earth on each individual atom in my body is unbelievably small, the effect adds up, so that the attraction of *all* the atoms in the Earth on each atom in my body is substantial (most noticeably in the morning, just after my alarm goes off). We don't "feel" the electromagnetic force in this way, because the negative charges in our body are exactly canceled by the positive charges in our body. As I suggested in chapter 4, this is a good thing; if it weren't so, the electric forces would explode us out of existence.

As weak as gravity is, we can still measure the gravitational attraction between human-scale objects. (The attraction between single atoms is so small that there's no hope of measuring it directly in the near future.) In fact, about 100 years after Newton's discovery of the law of gravity based on the motions of the planets around the Sun, a fellow Englishman, Henry Cavendish, came up with a sensitive method to measure the gravitational attraction between objects the size of cannonballs by attaching two to a crossbar to form a kind of dumbbell balance and suspending it from a wire. He then moved a third cannonball close to one end of this contraption and measured the infinitesimal torque this produced on the wire. In this way, the fundamental strength of gravity itself—the so-called gravitational constant—was determined. Previously, one could use Newton's law to calculate the strength of the gravitational force between planets and the Sun, or between the Earth and the Moon, for example. However, the mass of these objects was not independently known; one could not determine how strong the gravitational force was between objects of known mass in this way. After Cavendish's experiment, not only was this measurement possible, but one could put the gravitational constant into Newton's law and in this manner *weigh* the planets and the Sun. The current best measurement of the mass of the Sun is due to this technique.

The purpose of my discourse on gravity's weakness, however, is not to bury gravity but to praise it. There is nothing basically wrong with imagining a universe full of invisible things, some of which are beyond our control. The universe *is* full of invisible things, some of which are beyond our control! We should think about gravity whenever we ponder the Big Question that has stirred our imaginations for centuries (and inspired much of modern science fiction): What invisible things are still invisible?

At the top of the list, anyone's list, must be ESP. It's difficult to name a major work of science fiction or fantasy that does not somewhere contain an element of telepathy. Each of the *Star Trek* series, for example, has had its telepaths: Spock, Deanna Troi, her mother Lwaxana, Kes—to say nothing of a host of telepathic aliens on various planets. The aliens in *The X-Files* perform telepathic mind scans; and even the disgusting creatures in *Independence Day*, whose only purpose in life seemed to be to kill other species, used telepathy as a weapon.

How many times have you felt that you knew what someone else was thinking? Certainly, as we become accustomed to reading body language and facial expressions, we can sometimes anticipate other people's reactions, or even divine what is on their minds. Is it all that crazy to imagine that with one more step we could communicate without speech?

The term "extrasensory perception" was coined by the Duke University researcher Joseph Banks Rhine, who wrote a well-known book by this name in 1934 in which he claimed to have overwhelming evidence for telepathic communication. His popularizations, combined with the interest of the publisher of the pulp magazine *Astounding Science Fiction*, helped fuel public interest and inspired a raft of ESP-related science fiction. Rhine also coined the term "parapsychology," for the study of various kinds of alleged psychic phenomena.

Alas, the invention of these two serviceable terms was probably Rhine's greatest contribution to science, since essentially all of his ESP results that were subjected to outside scrutiny were shown to be flawed—including his first discovery, Lady Wonder, the telepathic horse. While the flawed experiments of one researcher cannot be used to dismiss a whole field, the following facts are not in dispute:

1: In the more than 60 years since Rhine created the field, there has been not a single definitive experiment broadly accepted—that is, by scientists not directly involved in similar lines of research—which unambiguously demonstrates the reality of any of the phenomena he set out to explore and promote.

2: At the same time, huge numbers of people, including a number of active workers in this field, believe that ESP exists.

I know better than to try and resolve this debate. Moreover, I have never personally tried to verify or debunk any specific set of ESP experiments. I'm skeptical, but then I try to be skeptical of everything (I don't believe there's any other way to learn about how the world really works). But I don't want to directly question here the quality of current research in this area. Rather, I want to ask a question I think is more enlightening, not to mention more fun: What would be required for ESP to exist?

I find it significant that the furor over telepathy and ESP began within a few decades of the invention of the radio by Guglielmo Marconi, and less than one decade after its first widespread usage. Once wireless communication became a reality, the idea that invisible "waves" of some sort could lead to direct nonverbal communication between people probably became a lot more plausible. Until then, the only nonverbal communication that didn't make use of some overt physical connection between source and receiver involved visible light, so that any suggestion that one might receive invisible signals was completely unprecedented. Radio waves fit the bill perfectly.

There are so many remarkable aspects of radio waves (which,

like visible light, are electromagnetic waves, but of much lower frequency), that it's hard to know where to begin talking about them. First and foremost, in spite of both the curvature of the Earth and the long distances involved, shortwave radio signals can be received on the other side of the planet. Moreover, though radio waves carry very little power, they can be precisely detected. The most striking illustration of this sensitivity is afforded by the marvelous Arecibo radio telescope in Puerto Rico. Built in a natural crater filled with tropical vegetation, the Arecibo antenna is 1,000 feet across, and viewers of the movie *Contact* will recognize it. It has detected radio waves from the surface of Venus, from rotating neutron stars thousands of light-years away, and from extragalactic objects hundreds of millions of light-years away. I toured the facility with the assistant director a while back along with my wife and daughter, and I remember trying to think of a way to convey how sensitive this beautiful device was. Based on the sensitivity data for the instrument, I worked out that it could easily detect a 25-watt lightbulb on Pluto, several billion miles away, if instead of generating visible light the bulb emitted its energy as a radio frequency accessible to the telescope's receivers.

Well, if we can detect such small sources located in the outer reaches of the solar system, why shouldn't two minds be able to communicate across a room? After all, thinking itself involves precisely the same processes as those that produce electromagnetic disturbances. Thoughts and actions are initiated by the firing of neurons in our brains, which produce electrical currents, which in turn travel to nerves and muscles elsewhere in our body. Electrical currents are precisely what generate electromagnetic waves.

On the surface, the forces of electricity and magnetism seem very different. Permanent magnets exist, but they behave quite differently than electric charges do. For example, if one cuts a magnet in half, one does not produce an isolated north pole and an isolated south pole; instead, one gets two smaller magnets, each with a north and south pole. But if I bisect an object with a positive electric charge on one side and a negative charge on the other, I will end up with one positively charged object and one negatively charged object. There is clearly some connection between electricity and magnetism, however. For example, I can create a magnet by moving charges to produce an electric current. These electromagnets are standard components in almost every electric appliance in your house.

Near the end of the nineteenth century, one of the greatest theoretical physicists of that era, the Scottish physicist James Clerk Maxwell, arrived at one of the greatest intellectual unification of ideas that has ever taken place on this planet. He demonstrated conclusively not only that electricity and magnetism were related but that they were really just different aspects of the same thing. One person's electricity is another person's magnetism, depending on the reference frame.

Besides setting the stage for relativity theory, which is based on this principle, Maxwell's theory of electromagnetism made a central prediction: Light is a wave of electricity and magnetism. The interplay between electricity and magnetism was such that whenever you jiggled an electric charge, a "wave" of electric and magnetic disturbances traveled outward at a speed that could be calculated from first principles. This speed turned out to be the same as the measured speed of light. We now understand that the frequency with which you jiggle the charge determines the measurable characteristics of the resulting wave. If you jiggle it back and forth only a million times per second, you will produce radio waves. If you jiggle it back and forth a billion times per second, you will produce microwaves. If you jiggle it back and forth a million billion times per second, you will produce visible light. And so on.

You might ask, what *is* it exactly that is propagating in an electromagnetic wave? What is there in the wave itself, and what will the wave do when it encounters matter? Here we have to thank another remarkable nineteenth-century British physicist, Michael Faraday. Faraday is in some ways a more romantic figure than Maxwell. Without a formal education, as a mere bookbinder's apprentice, he attended a public lecture in 1812 at the esteemed Royal Institution, in London, given by the brilliant chemist Sir Humphry Davy. Sometime later he returned to the institution with the lecture notes he had taken, bound into a handsome volume. Davy was so impressed that he took Faraday on as an assistant. The rest is history.

The particulars of this history involve a number of seminal discoveries about the connections between electricity and magnetism which set the stage for Maxwell's work. But the one I want to focus on here is one that changed forever the way physicists think about empty space. Faraday was an intuitive, seat-of-the-pants type of thinker, which is one reason I like him so much. Prior to Faraday, when physicists thought about forces, like gravity, they pictured the equations that governed these forces. Faraday provided a more intuitive, physical picture, which in some ways is far more valuable.

From the moment Newton discovered the universal law of gravity, he and others were puzzled by the question, How does the Moon know the Earth is there in order to be attracted by its gravitational pull? That is, what exactly is it that communicates the force of gravity? Is that force instantaneous, or does it take time to reach the Moon?

Newton never resolved these thorny questions, and preferred to move on to other things, including becoming head of the British Mint. Some 200 years later, however, Faraday pondered the same questions, but this time in the context of the electric forces between particles. To help himself understand why the electric force behaves the way it does, he imagined that emanating from every charged particle was an electric "field." He pictured this field as a set of lines radiating outward in space from the particle in every direction. If he imagined the number of lines as proportional to the magnitude of the electric charge on the particle, Faraday could then understand why the electric force dropped off in strength with the square of the distance between charged objects. If I start out with a certain number of field lines emanating from a charge, and each one heads out in a straight line to infinity, the field lines will diverge. Therefore, the number of field lines that cross any given area at a certain distance will decrease with the square of the distance.

Now, this is a nice picture, but is it more than just a metaphor? Often physicists create pictures to give themselves a clearer understanding of how the laws of nature work, but are these pictures ever the image of the reality itself? Sometimes the answer is a surprising yes. Faraday's fields are such an example, and soon took on a life of their own. It was shortly understood that under certain conditions electric and magnetic fields could be generated simply by the presence of other electric and magnetic fields, without the presence of the electric charges that caused one to invent the fields in the first place.

When physicists nowadays think of empty space—space devoid of matter—they realize that it's not necessarily empty. We now think of the electric force, and also the gravitational force, as follows: A charged particle creates an electric field around itself, and a massive particle creates a gravitational field around itself. These fields propagate at the speed of light, and a far distant object can interact with them and be attracted or repelled. Because it takes some time for the fields to propagate, the Moon, for example, will be gravitationally attracted to where the Earth was at the time the field with which the Moon is interacting was created. If the Earth moves in the meantime, the Moon will nevertheless move toward the original place—that is, until the field created by the now moved Earth propagates out to the position of the Moon. Because these fields propagate at the speed of light, we don't normally notice the delay on a human timescale. However, when cosmic distances are involved, the effects of the finite propagation speed of gravity can be dramatic. For example, the Milky Way is falling toward a huge galactic group some 50 million light-years away. In the time it has taken for the gravitational field of the huge cluster of galaxies to propagate to the region of our own galaxy, the cluster has moved from the position to which our galaxy is being attracted by perhaps 100,000 light-years, a distance comparable to the Milky Way's diameter!

Empty space is full of fields. A million years after I jiggle an electric charge here on Earth, the changing magnetic and electric fields have propagated a million light-years away, where they can cause an electric charge in an antenna attached to a radio receiver to jiggle up and down, producing a response in the receiver. The opening sequence of *Contact*, in which we pass slowly out through space, following the stream of electromagnetic waves emanating from our radio and TV broadcasts as they make their way through the universe, is a wonderful illustration of this idea.

We sense directly only a small part of all the electromagnetic waves out there. This spectrum includes waves with frequencies to which the electrons in the atoms in our eyes can respond, sending signals to our brain which we interpret as one or another color. Waves of slightly lower frequency are invisible to us, but we nevertheless feel them as heat. Waves of slightly higher frequency are invisible—to us, though not to, say, bees—and we don't feel them at all, but they damage our skin and produce dangerous but apparently appealing suntans.

What could be more New Age than this? An invisible world full of electromagnetic fields all around us, some of which we generate by our own thought processes. How cosmic . . . ! Why couldn't our thoughts generate weak fields that might be sensed by individuals with just the right kind of antennas built into their brains?

But this is a case of too much of a good thing. Electromagnetic fields are remarkably good at propagating and producing effects. But if they produce effects, they are by definition *observable*. That's the way the world works. If I think very hard—whatever that means—and try to produce a response in your mind, that means I must induce some chemical or electric response in the neurons in your brain. But unless you think your brain behaves differently from any other sort of antenna in the universe, then the signal I send to your brain should be detectable by radios or other types of electromagnetic receivers in the vicinity.

There's no doubt that the most sensible carrier for telepathic messages would be electromagnetic waves. There's no doubt that they are directly associated with the operation of your thought processes. We have detected "brain waves" and can even measure the external electromagnetic signal they produce. But electromagnetic waves from the other end of the universe are detectable by receivers here on Earth. Why should such receivers be less efficient at receiving telepathic messages than your brain is? The fact that no one has ever detected electromagnetic waves associated with ESP is pretty damning, don't you think?

Maybe the electromagnetic waves associated with telepathy are so weak that existing detectors are insensitive to them? But they can't be too weak to generate some physical disturbance in the brain of the recipient. This would entail carrying enough energy to cause an electron to jiggle, or an atomic spin to wobble, or something. But this same something can be used as the basis of some detection apparatus or other. Existing detectors of visible light can detect, for example, individual photons. We can build X-ray detectors to see through what we cannot see through with the naked eye, infrared-sensitive cameras to spy on our neighbors in the dark. The bottom line is that there is nothing more detectable in the universe than electromagnetic waves, as hidden as they seem.

No, this is another case where Fox Mulder's maxim, "The easiest explanation is also the most implausible," holds true. If ESP is to work, there's gotta be another way—something not quite so easy.

CHAPTER TEN

Mad, Bad, and Dangerous to Know

Thinking is very far from knowing.

—*Proverb*

Stare deeply into the eyes of someone you love, and you are sure to feel that you know what they are thinking. Their thoughts are as real to you as your own. Everything about this person seems tuned to your own visions and desires. You send out the signals, and wait.

Indeed, if you know someone well enough, you often *do* know what they are thinking! I recently had lunch with a physicist who said he read his daughter's mind on occasion. This statement surprised me considerably, but later on in our discussion it became clear that he really meant something more along the lines of what I stated above: He knew her so well that he often was able to anticipate what was going on in her mind.

Still, the lesson of the past century is that the universe is full of invisible fields—so many that Faraday himself would have been surprised. As you walk across the room, the number of invisible items impinging upon your body is staggering. Besides the complete spectrum of electromagnetic waves—the radio waves from nearby broadcasting stations or from distant galaxies, the infrared waves radiated by the heat of the walls or the bodies of other people in the room—we are bombarded by invisible neutrinos from the Big Bang, gravitational waves from collapsing stars in our galaxy, neutrons emitted by radioactive materials decaying in the ceiling and walls, not to mention the invisible Higgs field that many elementary particle physicists believe permeates space giving mass to all matter, or a possible invisible field associated with the mysterious "dark matter" that is thought to make up the greater part of the mass of the universe that I described earlier. As one gets to smaller and smaller scales, the presence of the various fields becomes more and more evident, so that on subatomic scales the elementary particles themselves can be thought of as manifestations of the fields which can create and destroy them.

There are a host of other phenomena out there as well, which, while invisible to the eye, can be detected by our other senses. A few molecules of perfume evaporating off the nape of a nearby female induce a flood of sensations and memories in your average male.

So who cares if electromagnetic fields don't fit the bill for ESP? The world seems full of senses and sensibilities beyond the five we know. If a bee can detect the invisible (to us) ultraviolet pattern on the petals of a flower, or a dog can hear the high-pitched squeal of a whistle in the distance while we hear nothing but silence, why cannot some of us detect at a distance the otherwise undetectable intense emotions of our loved ones, or even the more prosaic musings of our neighbors?

Extrasensory perception seems so palpable, so tempting, that it's hard to believe it doesn't exist. Psychologists and parapsychologists of varying degrees of eminence have advanced ideas with varying degrees of vagueness over the years. Carl Jung, in a leap of imagination unfettered by empirical evidence, posited the existence of a "collective unconscious" shared by all minds (not unlike the Borg of *Star Trek*). Others have argued that as humans developed language they lost the need for their innate ESP sense, much as our senses of hearing and smell, so essential to life in the wild, have been suppressed by our urban experience. Luther Boggs, a death row inmate on *The X-Files* who possesses ESP, goes a step beyond Jung and claims that "the dead, living . . . all souls are connected." Others have adopted jargon with a more scientific ring, like "morphogenetic fields," a term meant to describe energy emanating from all sources—animal, vegetable, and mineral—and carrying ESP signals.

Alas, the statement of Groucho Marx that he would not want to belong to any club that would have him as a member comes to mind. As noted in the last chapter, for a field to carry signals from one person's brain to another it has to (a) transport enough energy to make something happen, and (b) interact strongly enough so that in a brain's "ESP antenna" a signal can be received. One can imagine how both these things can be done, but such a field would not be undetectable by our present instruments.

We know of long-range fields in the universe, from electromagnetism, the strongest macroscopic field, to gravity, the weakest. Having dispensed with the former, let us work our way down to the latter. Let's consider a carrier somewhere in the range between gravity and electromagnetism: for example, the so-called weak force. The interactions mediated by this force between different particles in the nuclei of atoms are responsible for the reactions that power the solar furnace. This may not sound particularly "weak," but that's because while the reactions mediated by the weak force allow nuclei to change their identity, another nuclear force, called the strong force, is responsible for the large energies released when they do.

The weak force has an extremely short range (less than the size of a single atom) and therefore does not qualify as a direct carrier of ESP, but particles that interact only via this force may themselves travel long distances and thus transmit signals. It turns out that all known elementary particles in nature interact via forces stronger than the weak force, except one: the neutrino. For this reason, neutrinos are almost completely undetectable and can propagate over long distances unaffected. Neutrinos are streaming through your body as you read this. Over a trillion neutrinos from the Sun stream through your body at near the speed of light every single second of every single day. These solar neutrinos not only pass through your body without interacting with the matter in it, they pass through the whole planet without any appreciable interaction. In fact, they could pass through a billion billion Earths lined up in a row without any such interaction. In spite of their cosmic impotence, we have detected solar neutrinos by feats of technological prowess that few science fiction writers would have dared to propose—for example, we have noted the effect of the occasional rogue neutrino on a single chlorine atom in a tank containing 100,000 gallons of cleaning fluid. There remains the neutrino background from the birth of the universe, however, which no one yet has any idea how to detect.

In the earliest moments of the Big Bang, the temperatures and densities everywhere were incredibly high. At these levels of density and temperature (exceeding 10 billion degrees), even neutrinos could not sneak through matter unaffected. They would have remained in thermal equilibrium with the environment; if the surrounding gas was hot and dense, so too would have been the neutrinos. As the universe expanded and cooled, normal matter emerged—protons, neutrons, and electrons—and then formed atoms of the very lightest elements. Using straightforward calculations based on laboratory measurements of nuclear reactions, we have been able to predict that most of the protons and neutrons should have coalesced into the lightest element, hydrogen; about a quarter of them into the second lightest, helium; and a mere trace into the third lightest, lithium. And the cosmic abundance of these elements today fits the prediction: The universe is roughly 75 percent hydrogen and 25 percent helium, while the abundance of primordial lithium is only about 1 part in 10 billion. This agreement between theoretical prediction and observation is one of the triumphs of Bang theory and gives us confidence that another prediction of the theory—one that cannot be directly verified—is also true.

The same reactions that determine the ratio of protons to neutrons in the universe and explain the observed ratio of hydrogen to helium also suggest that a background of neutrinos from the Big Bang must exist, which permeates all of space. At any time, in a volume of material as big as a teaspoon, there should be roughly 100 neutrinos left over from the Big Bang. Like the solar neutrinos, these neutrinos are not actually sitting still in the teaspoon but streaming through it at or near the speed of light. Unlike solar neutrinos, however, these neutrinos carry a much smaller energy—a cosmic background neutrino has less than a millionth of the energy of a solar neutrino. Therefore, no one has ever been able to figure out a way to detect the neutrino background, even though physicists are persuaded of its existence. The discovery in the mid-1960s of the universal background of microwave radiation from the Big Bang—radiation produced by the same reactions that should have resulted in the invisible neutrino sea—gives us additional confidence that it exists.

So, here is a genuine candidate for a truly invisible background "field" that permeates the universe. But it gets better. Elementary particle physicists believe that other, even more weakly interacting particles might have been produced in the Big Bang. These particles are purely hypothetical, and they have strange names—neutralinos, axions, dilatons, and so on. Nevertheless, there are various fundamental puzzles about the nature of matter and the nature of the known interactions which can be solved only if such (as yet undetected) particles exist.

Better still, when we attempt to measure the total amount of matter in the universe—both within galaxies and between them—all indications are that there is far, far more than meets the eye. As I noted in chapter 6, over 90 percent of the mass of the universe seems to be invisible; it doesn't shine by emitting electromagnetic radiation. Could some exotic form of "dark matter" thus be the carrier of ESP signals?

No.

And neither could neutrinos, even the ghostly sort from the Big Bang. In fact, neutrinos illustrate perfectly the problems involved in positing any physical mechanism for ESP. For one thing, the fact that neutrinos are weakly interacting also means that they are very hard to produce. The processes that create them either happen very rarely or require enormously high energies. For example, neutrinos released in nuclear decays (such as solar neutrinos) generally carry energies over a million times higher than are carried by radio waves. This means, first of all, that if they did manage to interact in your body, the energy they deposited would be far greater than that required just to lightly jiggle an atom; rather, the energy deposited would be characteristic of other radioactivity and would not be particularly healthy. The cosmic neutrino background, on the other hand, is not energetic today because it has cooled considerably since it was produced. And the production of enough neutrinos via nuclear decays so that, say, 1 neutrino per second would interact with an atom in your brain, calls for a source at least 10 times as energetic as the Sun but contained in a volume the size of a breadbox and situated no more than about 1 foot from your head. I think any such neutrino-generated mental messages coming at you from, say, your lover or your dog would be irerelevant in this case!

These problems are even greater when it comes to particles more weakly interacting than neutrinos, like the purported dark-matter particles. One needs either the Big Bang or very-high-energy particle accelerators, like the 26-kilometer machine currently under construction at CERN (the U.S. Congress unfortunately canceled a more powerful machine, which was under construction in Texas), to produce very weakly interacting particles. To detect such particles one requires either very large detectors, occupying most of a city block, or else a tremendous amount of patience. I once worked out that even if one could build a detector capable of detecting a cosmic background of axions or neutralinos, the rate of energy deposited in the detector would be less than a millionth of a millionth of the energy produced by the residual radioactivity that exists in your big toe.

Alternatively, while there are some new types of elementary particles predicted to be produced in radioactive decays of ordinary matter, the (average) lifetimes before such decays occur are unimaginably long, because the forces involved are so weak. One current suggestion is that the constituents of normal matter, like diamonds, are not forever, and that all protons and neutrons in the universe will eventually decay, leaving nothing heavy enough to serve as the nuclei of atoms. But you don't have to worry about watching various parts of your body disappear. The predicted lifetime for such decays exceeds a million million billion times the age of the present universe. What is even more fantastical, I think, is that we can build—and, indeed, are building—large underground detectors that might be sensitive enough to catch one such rare decay.

So much for new weakly interacting particles. Any such new form of matter roped into service as a carrier of ESP will suffer from the Curse of Newton's Third Law: If you interact with me, I must interact with you. Neglect of this law is responsible for a ton of silly errors in science fiction. I discussed the infamous "ghost" error in my last book, wherein a ghost is seen to be too incorporeal to lift anything or to embrace its loved ones, but for some reason whenever it walks, it walks on the floor, and whenever it sits down on a chair, its butt manages to stay put. Here is another one: Every now and then in science fiction, including (frequently) in *Star Trek*, humans are briefly rendered incorporeal and can pass through walls, and so forth; sometimes this happens when they "inhabit" some other dimension, so that they don't interact with our 4-dimensional universe, and sometimes it happens just because they have been transformed into some form of noninteracting matter. However, in this case, how do they breathe? Presumably the oxygen in the air surrounding them, which is necessary for their survival, is as impervious to their existence as anything else.

There is rock left to turn over, though. We can't use the weak force because it is short-range, but what about some new long-range force in nature, beyond the four known fundamental forces. What about the fourth force, gravity, for that matter? It does, after all, literally make the world go round.

No one, to my knowledge, has suggested that gravity itself is the carrier of ESP signals—perhaps because gravity is so universal. Any old chunk of matter will do the job, and exactly the same job, apparently. There is nothing special about the gravitational properties of the brain, as far as we can tell. And even if there were, it's hard to see how even a brain as big as a dolphin's could produce a gravitational force on a nearby object big enough to do much of anything.

But is there a fifth force, so nearly invisible that we have so far been unable to detect it? This leads me to the pioneering work of the Hungarian baron Lóránt Eötvös. In Budapest in 1889, Eötvös began a series of remarkable experiments on the nature of gravity which he carried on over a 30-year period until his death in 1919. (Actually, his most famous paper appeared 3 years after his death, indicative of the pace of publication at the time and not of any afterlife experience.) The question Eötvös addressed is one that later formed the heart of Einstein's general theory of relativity: Does gravity attract all materials in the same way, regardless of their composition? If gravity represents the curvature of space itself, then clearly all objects should respond to this curvature in the same way. If they do not, then either general relativity is incorrect or a new force must exist beyond gravity which is sensitive to the composition of certain materials.

Now, you might think that since gravity itself is so weak, trying to distinguish small variations in the gravitational attraction between different materials would be impossible. However, if you are clever, it isn't. Eötvös performed an experiment in which plumb bobs made of various materials were allowed to hang freely, and the angle these made with the vertical were compared. Absent Earth's rotation, they would point directly down toward the Earth's center because of the force of gravity, but the rotation of the Earth pulls them a little off center. If the gravitational force on the plumb bobs was different because of the difference in the materials making up the plumb bobs, then the relative magnitude of the downward force and the sidewise force would be different, and the plumb bobs would make different angles with the vertical. By carefully comparing their angles, Eötvös claimed to put an upper limit on the difference in the gravitational force on different materials of less than 1 part in 100 million!

What does this mean? It means that if there *were* a fifth force that was material-dependent, its strength, compared to the strength of gravity, would be less than 1 part in 100 million! But things got worse.

In 1964, a modern experimental wizard, Robert Dicke (who helped develop the maser, the laser, the lock-in amplifier, the microwave radiometer, and atomic clocks, measured the solar oblateness, and devised a way of measuring the cosmic microwave background radiation from the Big Bang), performed an experiment using a sensitive balance and laser beams to measure the possible difference between the gravitational pull of the Sun on objects made of different materials. Dicke's experiments have put an upper limit of 1 part in a trillion on such a difference!

The experiment was so sensitive that the gravitational force on the balance due to the mass of an experimenter anywhere in the room would have produced an effect orders of magnitude bigger than the quoted upper limit. It was so sensitive that a fragment of iron 10-millionths of a meter on a side on either arm of the balance would have experienced a force in the Earth's magnetic field 100 times greater than the limit. A difference in temperature between the two balance arms of 1/10,000 of a degree would have ruined the sensitivity of the experiment. And so on. As a theoretical physicist, I am always in awe of such technological masterpieces, but the significance of these results from our point of view is that at scales for which the experimental sensitivity is the greatest, any new long-range force would have to be less than 1 billionth the strength of gravity. And I urge you to recall the arguments of the last chapter: Gravity is very very weak!

About 70 years after Eötvös's death, some desperate soul decided to reanalyze the baron's data and claimed to find evidence for a material-dependent force. This newfound fifth force dominated the physics literature for a few months. Of course, as any experimentalist will tell you, it is particularly worrisome to analyze the fine details of experimental data if you have not been a part of the experiment yourself—much less the details of data from an experiment performed 70 years earlier. The new "fifth force" quickly went the way of many other sensational but incorrect discoveries. Nevertheless, although the claim was disproved, it served a useful purpose. Many experimentalists realized that they had the technology to search for new forces that might act on a variety of scales, from meters to miles. Within a couple of years, experimental results began to appear constraining a new material-dependent force on various scales, always at levels well below the strength of gravity, the weakest known force in nature.

Now, it is true that no one has ever put two people, one thinking and one ready to receive, on a Dicke-type balance, but given the plethora of experiments, I find it difficult to believe that any new force strong enough to tweak the atoms in another person's mind across the room can have escaped the attention of all those researchers. As long as one subscribes to the notion that our brains are made of the same stuff as everything else, then much as we would love to read another's mind, there seems to be no light to read it by.

With all this talk about gravity and new weak forces between mind and matter, I can't help but return to an idea far older than ESP. The Force in *Star Wars*, which led off this discussion, is in fact much closer in spirit to a notion originally at the basis of astrology, which in turn had its roots in ancient Greece. Suspecting that four elements—air, earth, fire, and water—were not enough to keep the universe going, Greek natural philosophers decided that there must be something else out there. This material, dubbed the "quintessence" (fifth essence) by Aristotle, was the material of the heavens, permeating all things—the fundamental essence of creation. It was of course invisible, and for two millennia—until it was shown not to exist in an experiment by A. A. Michelson and Edward Morley at Case Western Reserve University—it went by its more familiar name: the aether.

The aether, being universal, connected the world of the stars with the world of humanity. This idea took on new life in the mystical world of ancient Alexandria. Here it was woven with various forms of Eastern mysticism into a new religion, astrology. The aether was the medium that linked the human drama to the regular motions of the planets around the Sun, and astrology explained what the planets were up there for. In yet another example of our profound ability to imagine ourselves at the center of the universe, the Alexandrian astrologers determined that the planets governed human affairs. The idea was not unique. In Rome, after all, the planets were gods, and mortals were subject to their caprice.

Twenty centuries ago, the notion that the aether should exist was a good enough hook on which to hang a new philosophy like astrology. However (except perhaps in the Reagan White House), that is not supposed to be good enough today: Astrology is neither internally consistent nor supported by experiment. (In my favorite example of the acuity of astrology, several people were provided with a horoscope, which was actually that of a famous serial killer. They mistook it as their own, confident of its characterization of their own personalities and experiences.) Still, most newspapers in this country carry a column on astrology, the annual sales of astrology books in the United States is about 20 million, and presidents and their wives find nothing particularly strange about determining their actions using the predictions of a "science" based on a material that was shown more than 100 years ago not to exist.

As myth gave way to science, the aether took on a more scientific cast. Newton and his contemporary the Dutch physicist and astronomer Christiaan Huygens had already established in the seventeenth century that light was a wave. This presented a problem, however, because a wave needs some medium to travel in. Remove the air surrounding a buzzer in a jar, and the sound disappears. There is no sound in space (as Gene Roddenberry knew but ignored). But what about light? The light that carries the image of the buzzer still reaches you even if the sound waves don't. There had to be some material, other than air, that light traveled in. What better choice than the aether?

So, from the seventeenth century to near the end of the nineteenth, the aether flourished for a scientific and not a mystical reason. In 1887, when Michelson and Morley demonstrated that there was no evidence for aether, they did not suspect that within 20 years a young theoretical physicist, Albert Einstein, would show that not only did experiment provide no evidence for the aether but theory now rendered the existence of such a material impossible. As far as science fiction writers were concerned, Einstein may have dispensed once and for all with the aether, but he gave them something much richer, a universe in which space and time themselves were relative.

One might say that the dark matter of modern cosmology is our modern-day aether, and in some general philosophical sense it is. It seems to permeate the universe, and it is (thus far, anyway) undetectable. There is one big difference, though: Dark matter is inconsequential on the human scale. It does nothing for us. Its gravitational attraction may determine the very expansion rate of the universe, but as far as day-to-day human activity is concerned, it may as well not exist. It is invisible *precisely* because it does not interact with stars or other forms of normal matter. It could no more tell us whether Mars was rising in the house of Aquarius than it could write a sonata.

Nevertheless, while astrology just seems silly, ESP remains a domain where one can continue to ask valid questions. While the data is discouraging—with the best candidate, electromagnetism, being too detectable and the handiest candidate (in terms of its ubiquity), gravity, being too weak—the search for new forces in nature is still an important enterprise. There is little doubt that undiscovered forces and undiscovered elementary particles exist at some level. While they play no direct role in our everyday lives, understanding them will inevitably help us to understand the processes that led to our own existence. This is the real "cosmic" connection to the stars provided by modern science, replacing the mystical renderings of astrology.

We are far from knowing everything about the nature of matter, and while there does not seem to be room to wiggle even a single undetected thought through the maze of modern experiments, the fact that we can test a force at a level of 1 part in a billion—a force so weak that it takes something the size of Earth to bring it to your attention—provides hope that if there are any strange new things that go bump in the night to be found, we will eventually find them.

CHAPTER
ELEVEN

It's About Time

SCULLY: Time can't just disappear! It's a universal invariant!

MULDER: Not in this zip code!

For a former physics student who was supposed to have done her undergraduate thesis on "A New Interpretation of Einstein's Twin Paradox," Dana Scully should have known better than to utter the above statement. Or maybe she had a *really* new interpretation. For at the very heart of Einstein's special theory of relativity—at least, the relativity I know and love—is the fact that time is precisely not a universal invariant. It flows at different rates for different people in different circumstances of motion or in different gravitational fields.

Now, what's Einstein got to do with ESP? Well, first off, back in the early heyday of ESP research, before the results of J. B. Rhine and others had been discredited, Einstein remarked that he had an open mind on the subject but would not believe it until he saw a "distance" effect. He was alluding to the fact that all long-range forces in nature fall off with distance, gravity and electromagnetism being the prime examples. Radio waves become less intense the farther away the source, so why shouldn't "brain waves?"

The fall-off with distance is really just a consequence of energy conservation. A source radiates a signal with a certain energy, and if that signal later spreads out over space, the energy per unit area carried by the signal has to decrease. There is no way around this. Not only has the conservation of energy been tested to the *n*th decimal place, but—more important, perhaps—we now understand that energy conservation is a consequence of the fact that the laws of nature don't change with time.

It seems to me that Hollywood has generally caught on to this sensible idea. In *Star Trek,* for example, Spock has to touch the person whose mind he is reading, and Deanna Troi has to be near someone in order to sense that person's feelings; even the perpetrator of ESP rape in the *Next Generation* episode "Violations" apparently had to be on the same spacecraft as his victims. The captured alien in *Independence Day* likewise seemed to have to be in the same room with its victims before it could kill them by psychic means. (If not, of course, why bother arriving here in spacecraft—why not just zap everyone from home?) And the spooky children in the sci-fi classic *Village of the Damned* could sense your malevolent thoughts only if you were nearby. Some of the wackier shows on TV have lately seemed to stray from this principle—but then, they're wacky.

In any case, following Einstein's remark, some ESP researchers searched for this distance effect and never found it. But of course they never found any irrefutable evidence of ESP itself, so there was no way to tell whether or not it fell off with distance.

More important, perhaps, than distance dependence is the relation between the ideas of time and space central to the theory of relativity, and ESP issues of clairvoyance and precognition. It is a tenet of relativity, one that has been verified over and over again, that no signal can travel faster than light. This means that instantaneous knowledge of remote events is simply impossible. Of course, the speed of light is pretty fast, so this isn't much of a limitation on terrestrial communication; however, it would rule out psychic messages from inhabitants of a distant galaxy who were not around 10 million years ago, if 10 million years is the amount of time it would take light to reach us from that galaxy.

More important still, this tenet of relativity defines the nature of time itself and also determines the nature of cause and effect, which is upended by some of the notions associated with ESP and precognition.

Back to Scully. One can plausibly argue that the flow of time is *not* invariant in nature by recognizing relativity theory's treatment of space and time as related aspects of the picture of the universe as 4-dimensional spacetime. Therefore, what happens to distance can happen to time. And everyone knows that the distance between New York and Boston depends upon the route you take. If you go due north on the Merritt Parkway and up through Hartford and then east on the Massachusetts Turnpike, the distance that elapses on your odometer will not be the same as it would be if you had taken Interstate 95 along the coast. But two people who travel these routes can still meet for a drink at the same place in space—say, on top of the Prudential building in downtown Boston.

Following this logic, why can't these two people take different routes between two points in spacetime, with more time having elapsed on one person's clock than on the other's? This is precisely what *does* happen, if one of them travels out on a fast rocket for a long time and then returns to Earth, while the other sits in her armchair reading this book while her friend is away. It goes without saying that for my reader the hours will fly like minutes and the days like hours—but when she eventually looks at her watch she will discover that the normal amount of time has passed. Her friend in the rocketship, who is traveling at a very high rate of speed, will have a watch, too, which will be running slow compared to the watch back on Earth. Thus—in this version of Einstein's classic twin paradox, studied by the undergraduate Dana Scully—for the rocket traveler the days that his friend passes on Earth *really can be* hours for him. When they meet again, she will be several days closer to his own age than she was when he left.

However, both these observers at least travel forward in time. As for traveling backward, Stephen Hawking made a telling argument for its impossibility: If it were possible, he said, we would be inundated by tourists from the future! I think this is a wonderful point (although I once countered it by suggesting that they all go back to the 1960s, where nobody would have noticed them).

Shirley MacLaine, too, has a take on the subject of time travel: In *Out on a Limb,* she writes, "In déjà vu you are getting an overlap of a past-life experience, or you could be getting an overlap of a future-life experience. . . . That's what Einstein said." Well, not quite, but Shirley does lead us to an important issue. If clairvoyance, precognition, or even some forms of ESP (communications from distant alien civilizations, for instance) were to exist, they would require a radical rethinking of what we mean by space and time—a rethinking that would perforce violate our other experience of the physical world. Seeing into the future means that in a sense the future has already happened. Moreover, "seeing" into the future means that somehow a signal from the future has "leaked" back in time.

While it isn't stated very often, clairvoyance and precognition present the same kind of paradoxes that the more pedestrian version of time travel does. The standard brain-twister is the grandmother paradox: Suppose you go back in time and kill your grandmother several years before your mother is born? Then you couldn't exist now. But if you don't exist now, how could you have traveled back in time in the first place?

Well, I can think of a version of this paradox which applies to clairvoyance and precognition. Say that you somehow intercept the future thoughts of your yet unborn great granddaughter, and that what you overhear alerts you not to marry the man you met on the bus today. Your future great granddaughter has apparently read, in some old family letters, that after he swept you off your feet and married you, he periodically beat you up. So, when he phones the next day to ask you out, you say no, and you never see him again. Therefore you have no offspring, or descendants, by him, so your great granddaughter by this man can't exist. But if she can't exist, how did you pick up her thoughts? The problems are as clear in the case of clairvoyance and precognition as they are in cases in which people are sent back in time from the future: If the Terminator had succeeded in his mission to assassinate Sarah Connor, then there would have been no need to have sent him back into the past in the first place; in *Back to the Future,* if McFly's nemesis had become rich by betting on sporting events whose outcomes he brought back from the future, then he never would have been a bum hanging out in the future near the time-travel car that brought him back to the past to start his financial empire.

Well, you say, that's simple. She was your great granddaughter, but not by this man! OK, if that's the case, then how does she know anything about his beating you? Uh-huh, so let's try another tack. Say that in the future she is your great granddaughter, but that the minute you say no to him, she will no longer exist, because the future has now changed. But from the point of view of the future, your "today" happened a long time ago. Hence, there is no sense in which *before* you overhear her thoughts and *after* you overhear her thoughts have any one-to-one correspondence to the times *before* and *after* your great granddaughter learns that your husband beat you. The events that immediately preceded and followed your phone conversation with the gentleman happened long before your great granddaughter was born, so how could her existence in the future change because of something that happened today? It would be a creepy world if people popped in and out of existence today because of something that happened many years ago.

We seem to feel that screwing up the future is more permissible than screwing up the past, but if you think about it the problems are essentially the same—especially if the future and past connect as they must in instances of precognition. In fact, seeing the future and traveling back in time are really equivalent, in a sense, since for you to see some event that happens in the future, some signal had to travel back in time.

The real issue here is cause and effect. A sensible universe, describable by physical laws, is one in which causes always precede effects, and not vice versa. Thus, even as real-world physicists consider whether or not a universe involving time travel is physically possible, they are careful to check that cause and effect are maintained.

For example, if time travel is possible (through spacetime wormholes, say), then a *round-trip* is possible. This means that one can potentially relive scenes over and over again. There are some episodes of *Star Trek* in which this happens. The crucial question is, Are you the same person you were the first time you lived through the experience, or can you remember the previous cycles? If you can remember the previous cycles, then you can change your actions, which is what happens in the *Next Generation* episode "Cause and Effect," when the *Enterprise-D* finally manages to escape collision with the *Bozeman*. However, this means that cause and effect go out the window, since you will have learned something from an event that has not yet happened to you. If cause and effect go out the window, then the laws of physics—every one of which, even quantum mechanics, is based on *causality*—would have to be modified. This is a lot to ask for a little precognition.

Instead, one might imagine a world in which even if time travel were possible, causality still obtained and future events could not influence past ones. Just as in the case of those who do not learn from history and are thus doomed to repeat it, in such a universe a "closed timelike curve" continually rehearses the same events, over and over, so that you cannot change the past no matter how hard you try. For example, if you go back in time to try to kill Hitler before he became Führer—when in fact he survived until shortly before the end of the Second World War—you will trip at the crucial moment, or the gun will misfire. In spite of its various psychedelic flights, the time-travel movie *12 Monkeys* honors this sacred principle. At one point, the Bruce Willis character, who has traveled back in time, tells someone that there's no need to worry about his presence in the time frame, because a time traveler cannot change history. Indeed, the Willis character's death in the movie is a classic example of a closed timelike curve. (This is not to excuse the film's incongruities, but I am not prepared to set to work on the *The Physics of 12 Monkeys*.)

In a world that contained such closed timelike curves, you could not change the past, and by the same token you could not change the future either. But this is a pretty boring view of time travel, and various individuals have tried to refute it in order to allow a more interesting and dramatic universe. (I like to imagine one of them as an expert witness in the O. J. Simpson trial, to bolster the defense that the real killers were time travelers, with Johnnie Cochran chanting, "If you can travel in time, there was no crime!") Anyone who seriously believes in precognition has to face these same issues: (1) How does the informing signal travel back in time? and (2) How does one deal with cause and effect?

Actually, I introduced the subject of time travel in the first place in order to apply a Hawking-like argument to the linked issues of ESP, clairvoyance, and precognition. It seems to me that the strongest constraint on such phenomena proceeds from the fact that Bill Gates is the richest man in America. I will now set out the argument, in case the connection isn't obvious.

Consider the following actual instance of a claimed positive ESP test result. In this experiment, a subject was able to correctly guess which one of 5 patterned cards was at the top of a face-down pile of such cards 947 out of 4050 times. Because there are 5 possible choices, one might expect a random guess to be right ⅕ of the time, leading to an expected number of correct choices of 4050/5 = 810. Probability theory tells you that this difference of 947 − 810 = 137 hits above the expected value would happen by chance only once in every 20 million repetitions of the experiment. Later on, the experiment was discredited, but that is not what I want to dwell on here; the point is that even ESP candidates don't come up with the right answer 100 percent of the time.

But let us say that, on average, a candidate does succeed in getting things right 10 percent more frequently than the laws of chance say he or she should. In that case, the following experiment can be performed. Have the candidate designate the 100 stocks which have the best chance of going up the next day. Random chance suggests that if the market remains on the whole static, then on average 50 of the selected stocks will go up and 50 will go down. But a good ESP candidate may skew these odds to 55–45. Therefore, as long as the stock market as a whole doesn't move downward, you are guaranteed to make money each day. Say that you increase your net worth by 1 percent each day in this way. In 5 years, if you started out with an initial investment of $100, then you will end up with $7,700,291,200.

One can certainly quibble with the details in the above example, but the point is that even a slight ability to beat the odds in any human activity puts you at a tremendous advantage in life. The fact that we don't often see instances of such exaggerated and apparently effortless gain is no more a proof against ESP and precognition than Hawking's argument is against time travel to the past. But it strongly suggests that either (a) there is no such thing as ESP, clairvoyance, and precognition, or (b) all those people who possess these faculties are keeping it and their money a secret.

CHAPTER

TWELVE

All Good Things . . .

I'm strong to the finitsch, 'cause I eats me spinach.

—*Popeye the Sailor Man*

I very much enjoyed the recent John Travolta movie *Phenomenon*, until about two-thirds of the way through. In the movie, Travolta plays a likable small-town innocent who is suddenly catapulted onto a new plane of existence after seeing a flash in the sky. He finds his mental capacities increasing each day thereafter: he can learn foreign languages in hours, read books in minutes, and so on. His reactions, and the reactions of those around him, to these newfound powers are entertaining, and the mystery of why this has happened is intriguing. Eventually, he goes into the office of his country doctor to be examined. Needless to say, the doctor is incredulous until Travolta provides a demonstration. Pointing to a pen on the doctor's desk, Travolta concentrates . . . until the pen slides across the desk into his hands!

It seems that whenever we imagine superintelligent beings, one of the first attributes we provide them with is mind over matter, the ability to control inanimate objects. Again, this seems a sensible extrapolation from our own experience: There is little doubt that our minds can control our own matter. By this, I don't simply mean that our brain dictates the movements of our body, but that we can also apparently control even subtle aspects of our physiology—our heart rate, our blood pressure, our pain threshold, even sometimes our recovery rate from illness. Could not a superior intelligence therefore control matters *outside* the body in which it was confined?

Of course, it is not always superior intelligences who are able to perform telekinesis. One of my favorite examples occurs in the original *Star Trek* episode "Plato's Stepchildren." The residents of the planet Platonius, an unpleasant society based on the teachings of Plato, developed telekinetic powers after ingesting native plants containing "kironide," a rare and potent chemical compound—that is, all but Alexander, who, because of a pituitary hormone deficiency, was unable to absorb kironide. Since he has no telekinetic powers, Alexander is at the mercy of the other Platonians, and has been forced to act as the court jester. Kirk and the rest of the crew are lured to the planet by a distress call, and are likewise forced to perform for the Platonians. Whatever the chemical weakness of the plot, at least one point is well taken: kironide or no kironide, something has to provide power for the process of telekinesis.

It may surprise you to find how much power is in fact needed. Say I want to do something simple, like lift a pen up from a table, or drag it across the table to my waiting hands. If the pen weighs, say, 4 ounces, or about 0.1 kilograms, then the energy required to lift it a few feet off the table or drag it toward you is about 1 joule. To expend this much energy in, say, 1 second means that 1 watt of power must be expended on the pen. Now, this doesn't seem like a lot, but if the pen is pointed toward you, it presents a cross-sectional area of about 1 square centimeter. If it is located 1 meter away, then this target represents only about 1/100,000 of the surface area of a sphere with a radius of 1 meter. This means that if you were to radiate a signal that moves out uniformly in all directions, then in order for 1 watt of power to be expended on the pen, you would have to expend 100 kilowatts of power—more than the kilowatt output of most big-city radio stations!

Of course, if you were able to beam your signal, like a laser, then the power requirements would be reduced, but given the geometry of your head (if that is the source of your signal), it is unlikely that you could reduce the spread of the signal by more than, say, a factor of 100 or 1,000. Thus at least 100 watts of power would still have to be generated and projected. This task would be, as we physicists like to say, nontrivial. One hundred watts, for example, is comparable to the total amount of energy generated in your body as you are performing your normal daily activity.

However, once again science fiction points in a possible direction. When Obi Wan Kenobi exhorts Luke to "use the Force," he is clearly implying that there exists some energy available in the ambient space and that you can learn how to tap into it. And Obi Wan is not alone. Many a sci-fi theme is based on tapping the resources of empty space. Now, this may sound ludicrous. How can empty space contain energy to be tapped? But in fact, the issue of whether or not empty space possesses energy is one of the most important in all of modern physics.

A fascinating realization to come out of twentieth-century physics is that empty space—and by this I mean *really* empty space, devoid not only of matter but also of any radiation, such as radio waves and so forth—is not empty. The central law of quantum mechanics (which I argued in my last book makes Gene Roddenberry's transporter impossible) is Heisenberg's uncertainty principle. When coupled with the special theory of relativity, it implies that empty space is full of what one might call "virtual reality." The uncertainty principle tells us that because of correlations between various sets of physical observables such as position and momentum, or energy and time—correlations that appear only at the quantum level and not in the macroscopic classical world—it is impossible ever to measure both members of these sets beyond a certain level of certainty. Thus, it is impossible to simultaneously measure both the position and momentum of a particle exactly. Likewise, if one measures a system over a certain time interval, one can never pin down its energy exactly: to do so would require measuring the system for an infinitely long time.

Einstein's special theory of relativity tells us that systems that can be measured must travel at speeds less than the speed of light. Since any given system's clocks will slow down as the system approaches light speed, one can show mathematically that if a system exceeds light speed its clocks will travel *backward*. Even though there is no reason to believe in the existence of superluminal objects, they have a name—tachyons, which is liberally used on *Star Trek* (various beings emit them, and they are associated with cloaked Romulan vessels).

If we combine Heisenberg's uncertainty principle with relativity, the two together imply that empty space need not be empty. The reasoning—which, again, I borrow from Richard Feynman—is straightforward, if somewhat wild. First, the uncertainty principle tells us that over very short distances and times, since we cannot measure the momentum of a particle exactly, there is nothing that stops it from traveling momentarily faster than the speed of light. But if it does do so, then it must behave like it is traveling backward in time. But if it behaves like it is traveling backward in time, then it must pass by its former self traveling forward in time. If it then slows down and starts to travel forward in time again, it will pass its intermediate "backward-in-time" self. This means that if I start out with one particle, for a brief time three (almost) identical particles will coexist: (1) the original particle, (2) the original particle traveling back in time to get there, and (3) the original particle which, after traveling back in time, slows down and travels forward again.

I insert the parenthetical "almost" because it turns out that if the original particle had an electric charge, then when it travels back in time it simply behaves like a particle with the *opposite* electric charge traveling forward in time. Thus, if one starts out with a proton, for a moment one will have two protons (the two particles traveling forward in time) and one antiproton. Indeed, it is precisely this reasoning that led some 70 years ago to the prediction that antiparticles must exist. After a while, you will end up with just one proton again. To an observer traveling forward in time (if you *could* observe the particles directly, which of course you can't, because if you could then the particle couldn't be traveling faster than the speed of light and thus backward in time in the first place), it would appear that you started with one proton and then momentarily a proton-antiproton pair appeared from nowhere only to disappear again shortly thereafter.

Now, this whole scenario seems too fantastical to be true, but it is. "How do we know?" you may ask—since by definition this process is supposed to be invisible. Well, while we can't see the three particles directly, we can indirectly detect their presence. For example, the electric field generated by three particles will be slightly different from that generated by one, even though the total charge will be the same. (The proton-antiproton pair has zero total charge.) So if the proton in question happens to be the nucleus of a hydrogen atom, the electron orbiting the proton will be influenced by a slightly different electric field than it would otherwise, and the energy levels of the orbiting electron will be altered in a calculable way. It was one of the great postwar successes of elementary-particle physics that this small shift could be both calculated and measured. Sure enough, the predicted effect of these particle-antiparticle pairs—called virtual particles, because they cannot be directly seen—is exactly in agreement, out to better than 9 decimal places, with the experimentally measured value.

The fact that empty space is filled with a boiling mess of virtual particle-antiparticle pairs that spontaneously appear and then disappear before you can say "Rumpelstiltskin" suggests that empty space may, in the quantum world, actually carry energy. In fact, in general it should. We know, for example, that as the universe evolved and cooled, the energy of empty space—or the vacuum, as it is usually called—changed as the temperature changed, and various so-called phase transitions took place to make the universe look the way it does now. So it is natural to suspect that even today the vacuum carries significant energy.

Which is a puzzle. If empty space carries significant energy, then this energy would produce gravitational effects that would change the way the universe expands. But observations of the expansion put an incredibly tight upper limit on the amount of energy that can be associated with empty space—an unbelievably tight limit, in fact: a billion billion billion billion billion billion billion billion billion billion billion billion billion billion times smaller than most theoretical physicists would have predicted, given our ideas from fundamental particle physics.

This puzzle—why empty space doesn't have a whole lot more energy associated with it than the observations allow—has become known as the Cosmological Constant Problem, and it is probably the most severe numerical conundrum in all of physics. There is no doubt that trying to resolve it will take us to the very deepest understanding of the fundamental laws governing the universe.

The name for this vacuum energy—the cosmological constant—refers to a theoretical speculation of Albert Einstein's, which he later discarded. In 1916, when he was developing his general theory of relativity, he realized that if it was to be a theory of gravity, it should be applicable to the universe as a whole. Current wisdom suggested that the universe was static; however, there was no solution of the general relativity equations that allowed for a static universe, if normal matter was all there was. The reason is simple. Normal matter attracts other matter gravitationally. If you lay out matter randomly in the form of stars and galaxies at rest throughout the universe, slowly and inexorably the gravitational attraction between these systems will begin to cause the whole system to collapse inward. Einstein soon realized that a way out of this problem was to add a term to his equations, the cosmological constant, which represented a kind of cosmic repulsion between matter on large scales. By balancing this repulsion with the standard gravitational attraction, one could arrive at a static solution of Einstein's equations—a solution that Einstein hoped would describe the universe we live in.

Within little more than a decade of this proposal, however, Edwin Hubble and others had convincingly demonstrated that the universe was expanding. In an expanding universe, there is no need for a cosmological constant, because attractive gravity can simply slow the expansion. As soon as Einstein became aware of the expansion, he dispensed with the cosmological constant, calling it one of his worst theoretical blunders. The only problem is that we now realize it was not his to dispense with. The vacuum energy associated with the virtual particles I have described would produce exactly the kind of term that Einstein added by hand to his equations. So the problem becomes trying to understand why this term in Einstein's equations, which we now understand should be there, must be (because of observation) 125 orders of magnitude smaller than arguments based on particle physics suggest.

Now, you might say, and some physicists do, that since the upper limit on the allowed energy is so small, why not just assume that the actual value is zero, and that some as-yet-unknown law of physics sets it so. This may be the solution to the problem. However, no one to date has come up with a convincing argument as to why this energy in the vacuum should be zero. Moreover—and more significant to me, at least—there is growing cosmological evidence that perhaps the energy density of the vacuum is not exactly zero. Along with my colleague Michael Turner at the University of Chicago, I have been championing this once completely heretical idea for over a decade. Who knows? It might even be true. If it is, a number of fundamental puzzles in modern cosmology might be resolved.

But although we might resolve cosmological paradoxes, we will have raised new problems for particle physics. It might be that no one has any really good understanding of why the cosmological constant should be zero, but at least one can imagine plausible new physics arguments for its being zero. If the cosmological constant is instead very small, we all will have a lot of homework to do.

As I mentioned earlier, while preparing this book I queried a group of the most prominent theoretical physicists working in particle theory and general relativity as follows: If there were one question about the universe you could receive an answer to, what would that be? The responses were remarkable in their variety and depth. One of the people I contacted was Edward Witten, a brilliant mathematical physicist currently at the Institute for Advanced Study, in Princeton. Witten works on string theory, an area of physics originally designed to address the fundamental paradoxes arising when one tries to reconcile quantum mechanics and gravity, but which many people hope will provide a unified theory underlying all the known forces in nature. His answer caught me by surprise. I had expected he might want to know if string theory described the real world; instead, he indicated that he would like to know if the cosmological constant was zero, and if it was, why, and if it wasn't, why not. In retrospect, this is understandable: any Theory of Everything worth its salt must address this fundamental issue.

Be that as it may, let us finally return to the issue at hand for John Travolta and Luke Skywalker. Could the vacuum, if it indeed carries energy, and if that energy were properly tapped, provide a power source of the type envisaged by old Obi Wan? I had never thought of the Cosmological Constant Problem in this context before now, but one can do a simple estimate, based on the maximum allowed energy which could be stored in the vacuum as a cosmological constant consistent with the observed expansion of the universe. If one were somehow able to release the energy stored in 1 cubic meter of the vacuum, it would amount to 1 ten-billionth of a joule.

This allows me a brief jab at the Transcendental Meditation movement of the Maharishi Mahesh Yogi. This group, which started out as a rather innocuous bunch claiming that meditation could help you feel and work better (probably true), has proceeded over the years to increase its promises. Now it is claimed that not only can TM help you "fly" momentarily but it can also help slow the aging process, and if enough people perform it, it can reduce the crime rate. I certainly believe that if a fair fraction of the Earth's population meditated regularly, the crime rate would decrease, because presumably a portion of the criminal population would be meditating some of the time instead of preying on the public. But flying is another matter. The TM movement is the only group I know of that pins its claims so thoroughly on modern physics. TM literature is full of explanations couched in the jargon of string theory and quantum mechanics. A theoretical physicist I have known since his student days is now head of the Physics Department at Maharishi University, in Iowa, and is a leading adviser to the Maharishi himself. On the side, he has run twice as a candidate for president of the United States, on the Natural Law ticket.

In any case, I have read somewhere the claim that it is by tapping the energy from the vacuum of the universe that TM devotees can momentarily fly. Using the maximum-vacuum-energy estimate above, I have calculated that to momentarily raise the

Maharishi a meter off the ground would require tapping into a cubic volume larger on each side than the island of Manhattan. (Perhaps they should call it the Unnatural Law Party?)

To lift in this way the pen that began this discussion is not much easier, requiring us to tap a mere 10 billion cubic meters of vacuum, or the space inside a cube 3 kilometers on a side.

The Force may be with us, all right—but don't hold your breath!

CHAPTER THIRTEEN

The Measure of a Man

There once was a man who said, "Damn,
It is borne in upon me I am
  An engine that moves
  In predestinate grooves,
I'm not even a bus, I'm a tram."

—*Maurice Evan Hare*

In spite of popular notions to the contrary, art and science will be forever intertwined. The limerick above was penned in the year 1905. In the popular consciousness, this was the dawn of a new century of progress and material success, based on the nineteenth-century mechanistic ideal. A well-managed world would run like a well-oiled clock. To many a scholar, and poet, the universe was a cosmic game of billiards, initiated by a master billiards shark and continuing indefinitely on its own, wherein the trajectory of history was as predetermined as the trajectory of the balls on the cosmic table.

This picture of the universe could not have been farther off the mark. That year would witness the birth of the two great revolutions in twentieth-century science, relativity and quantum mechanics—revolutions that would forever change the way we think about the world and our place in it. Paralleling these developments, within a generation the world would live through the unraveling of the great nineteenth-century European order, a World War, and a Great Depression.

As a new millennium approaches, the world is a much more uncertain place than it was at the last turn of the century. For one thing, our experience with the physical world at the various extremes of scale has taught us to expect the unexpected.

How is this change reflected in our literature—in particular, in our science fiction? Maurice Hare's tram has been replaced by the likes of HAL, the overzealous computer in *2001: A Space Odyssey,* and Data, the near-human android in *Star Trek*, and *The X-Files'* homicidal computer COS. The question no longer seems to be "How much like a predestinate machine is Man?" but "How much like a human being can a machine be?"

I am writing this shortly after a watershed moment in the history of the man-machine debate. The IBM computer Deep Blue recently defeated world (human) chess champion Gary Kasparov, marking the first time the best human chess player on the planet has been defeated in a tournament series by a computer. This defeat was particularly notable because Kasparov had predicted publicly before the contest that a computer would never beat a human world champion. After his defeat, he suggested to reporters that the machine showed signs of "intelligence." This highly publicized matchup has spawned a batch of articles in the popular press addressing the question: Can computers think?

This is not the first time, of course, that this question has been raised. Since digital computers first appeared on the scene, people have been wondering whether they might possess attributes heretofore thought of as exclusively human. The logician and computer scientist Alan Turing laid out the issues in a 1950 essay titled "Can a Machine Think?" and *Star Trek*'s Data has pondered them in more than one episode, particularly after his emotion chip was installed.

Each time a computer has crossed a new threshold, disproving yet another claim that "a machine will never be able to *X*," the debate has been reinvigorated. To some, the suggestion that computers may one day develop consciousness is heretical. These people link the concept of consciousness to their belief in the existence of a human soul, a nonmaterial entity supposed to embody our intellectual, emotional, and moral being—and thus, presumably, our consciousness. In many religions, the soul is viewed as immutable and indestructible, continuing to exist long after our material bodies have turned to dust. I have always had a hard time with this logic, since one's consciousness, or self-awareness—and thus, it would seem, one's soul—develops gradually after birth (or, if you are a stickler for embryonic rights, after conception). If a consciousness can be created where none existed before, then why should it not die with the body? (This is where the *Enterprise*'s transporter, as I indicated in my last book, would come in handy: If every atomic configuration in your body can be transported to another locale, and you end up as the same person, that circumstance would appear to dispense with the idea that a nonmaterial soul perfuses the body. That is, of course, unless you imagine that the soul can find the body wherever it may be located in space; that might explain why there are so many souls in *Star Trek* that get disconnected from their bodies only to find their way home eventually.)

In fact, there are a number of ways to wiggle out of the immutable-individual-soul idea. For example, one can invoke the notion of reincarnation, wherein the soul exists before birth. There is a huge numerical anomaly, however: there are more people alive at the present time than have been alive in the planet's previous history, so where have all the extra souls come from?

Well, one can argue—as do some religions and at least one *X-Files* episode—that some souls migrate from animals to humans, thus taking up the slack. But some might find this idea more offensive than the idea that computers possess a consciousness and with it a soul. Aside from that consideration, what about evolution? What about when all that was here on Earth was algae? Do algae have souls? Well, I suppose one can get around this by appealing to the cosmos—that is, perhaps our souls have transmigrated from others long dead, in other solar systems? One now must suspend disbelief about how the souls got here and wonder whether there is some cosmic Law of Spiritual Conservation, so that the number of souls present in the universe at any one time is constant.

Or perhaps one can appeal to a kind of "collective consciousness," in which all our souls are part of one coherent whole that exists everywhere at once and can be divided up into as many pieces as necessary. Avoiding the issue of where this reservoir would actually exist and how such an arrangement would be made consistent with a causal universe, we must conclude that a collective consciousness would certainly allow for such phenomena as ESP, channeling, and so forth—and it has a nice New Age flair to it. But, as I discussed in chapters 9 and 10, this requires one to believe that the mechanism of consciousness is nonphysical, since there appears to be no physical mechanism to mediate ESP in a way not easily detectable. However, the fact that the actual process of thinking can be observed using sensitive magnetometers suggests that at least some aspects of conscious thought—and thus perhaps consciousness itself—are physical.

One may then appeal to the last refuge (literally) of religion—namely, that souls reside in Heaven, a place inaccessible on a human plane—and hold that, like God and Heaven, the soul exists beyond physical law and cannot even be discussed in terms of physical law. There's no argument to be posed against this point of view, because it is deliberately untestable; one must accept it or reject it on the basis of faith. But it's worth emphasizing that reliance on faith is probably the only way to avoid the various logical pitfalls that confront the advocates of an immutable soul grafted onto human consciousness.

Now, if we can demonstrate that the source of consciousness is completely biophysical, is that the end of the human soul? No, I expect not. Fundamental religious tenets must evolve as science evolves, in order to remain viable. When the Earth was shown not to be the center of the universe, the Catholic Church survived the blow and moved on. Faith is not easily shed. Fox Mulder's motto "I want to believe!" applies just as well to conventional religion as it does to UFOlogy. I suspect that once we understand the physiological basis of consciousness, the theological realm of the soul will retreat, to avoid conflict with experiment.

Even for some of those who believe that the mind is physical, it is not easy to accept the notion that a computer might one day think in just the way a human being does. The mathematical physicist Roger Penrose is one of the most prominent believers in a fundamental irrevocable *physical* difference between man and machine. Penrose is convinced that digital computers can never achieve human intelligence and self-awareness. He has written at least two books on the subject, with slightly different premises. Lest what I am about to say be misunderstood, let me affirm the fact that Penrose is a far more brilliant mathematician than I am, and that he has clearly thought about and researched this issue at much greater length than I have. Many of his descriptions of modern physics are incisive and beautiful, but I find his arguments on the subject of human vs. machine intelligence completely unconvincing. The premise of his first book on the subject, *The Emperor's New Mind*, is that some as yet undiscovered law of physics operating in the shadowy realm where quantum mechanics and gravity meet, differentiates between the processes of human intelligence and those of a digital computer.

Most physicists think that what happens on this minuscule scale is completely irrelevant to an understanding of what goes on at the scale of the human brain—or even at the atomic scales relevant to chemical processes, which are themselves orders of magnitude larger than the scale where quantum effects become significant in gravity.

Penrose has slightly modified—or, at least, clarified—his arguments somewhat in his next book. Here he makes it clear that he believes that computational machines cannot think as humans do because of the mathematical theorems (proposed by, among others, Kurt Gödel and Alan Turing) proving that computational systems are necessarily incomplete. In other words, there are certain assertions that are true but can never be proved true within the context of any particular system of mathematical or computational logical rules. Since humans can understand the truth of such assertions through human intuition and insight, then human intuition and insight cannot be reduced to any set of rules. And therefore human understanding (read "consciousness" or "self-awareness") can never be replicated by computing machines.

It is interesting that Turing himself earlier rejected this argument as a basis for believing that machines cannot, in principle, think. He argued that there was no proof that similar limitations didn't exist for the human intellect, and further that a human being could triumph only over one machine at a time, not simultaneously over all machines. "In short, then," he wrote, "there might be men cleverer than any given machine, but then again there might be other machines cleverer again, and so on."

I find Turing's essay on the issue of machine intelligence, although it is almost 50 years old, a clear and most refreshing discussion. In this essay Turing proposed what has since become known as the Turing Test for machine intelligence. In the spirit of a physicist, it is an operational test, which Turing dubbed "the imitation game." If the machine passed—that is, if it most of the time succeeded in fooling a human interrogator, removed from it in another room, into thinking it was human—then the answer to the question "Can machines think?" would be in the affirmative.

Turing made his own views on the issue quite clear:

> I believe that in about fifty years' time it will be possible to programme computers, with a storage capacity of about 10^9 [bits], to make them play the imitation game so well that an average interrogator will not have more than a 70 per cent chance of making the right identification after 5 minutes of questioning. The original question, "Can Machines Think?" I believe to be too meaningless to deserve discussion. Nevertheless I believe that at the end of the century the use of words and general educated opinion will have altered so much that one will be able to speak of machines thinking without expecting to be contradicted.

Predictions of "educated opinion" are notoriously chancy, and Turing was clearly overoptimistic. While we now possess machines of the storage capacity he hoped for, I don't think that any of them have yet unambiguously passed the Turing Test (notwithstanding Gary Kasparov's suspicions that some of Deep Blue's moves were made by its human programmers). It is certainly clear to me that neither of the two most famous fictional intelligent computers, HAL and Data, would be likely to pass the test. (Despite this, I sided with Jean-Luc Picard when he argued before a Federation tribunal, in "The Measure of a Man," that Data was a sentient living being, entitled to the rights of such, and not merely property of the Federation.) Moreover, the issue of machine intelligence is still as hotly debated today as it was when Turing made his predictions.

A fundamental difference between Turing's arguments and Penrose's is that Penrose goes further than the use of mathematics to argue his point. He attempts to isolate the fundamental physical difference. From my point of view, this approach is the only reason that the issues of intelligence and consciousness are appropriate for physicists to debate (and the chief reason I have introduced them at all). As I understand it, Penrose claims that the difference between human intelligence and computing algorithms originates in the mysterious nature of quantum mechanics, which of course governs the functioning of the fundamental atomic constituents of the brain. Moreover, he argues that a full understanding of human consciousness will rely on new laws of physics, which he claims are required in order for us to properly understand how the classical world arises from the quantum-mechanical world. He introduces the unfortunate notion that a proper understanding of quantum gravity will be integral to this understanding of consciousness. However, even if one completely disagrees with this last claim, one can explore whether the non-classical physics associated with the human mind will distinguish it forever from a computer. And I believe that some exciting developments in the past few years suggest that the opposite is true!

As computers get smaller and smaller, the individual logic units—the "bits" of the machine—will eventually become the size of atoms. (Data's positronic brain apparently uses positrons, the antiparticles of electrons, but what the heck.) Richard Feynman used to speculate on how small you could make various machines and still have them work. He realized that once bits were the size of atoms, the laws of quantum mechanics, which allow atoms to behave very differently from billiard balls, must be taken into account.

Indeed, while the field of computer science is based on the mathematical theory of computation, computations are carried out using physical devices, and thus it is the task of physics in the end to determine what is practically computable and how. Since the physical world at a fundamental level is quantum mechanical in nature, the theory of computation must also take into account quantum mechanics. Thus, the classical theories of Turing and others on computation should really be thought of as approximations of a more general "quantum theory of computing."

It has been explicitly demonstrated in the past few years that many of the limits on practical computing with digital computers—which use standard, classical bits for their computations—can be overcome by quantum computers. Algorithms can be developed which, if the computer components are quantum mechanical in nature, will allow calculations to be made exponentially faster than classical computation theory allows. A particular example involves an algorithm to find a nontrivial factor of a large number (that is, a factor not equal to 1 or the number), but the specifics of this example are not important here. What is important is an appreciation of why quantum computations can be different from classical ones. But to get an inkling of some of the physical processes that may well underlie our conscious awareness requires us to enter the murky world of quantum mechanics and explore phenomena that defy all classical reasoning.

CHAPTER FOURTEEN

The Ghost in the Machine

> After three hours I asked him to summon up the soul of Jimi Hendrix and requested "All Along the Watchtower." You know, the guy's been dead twenty years, but he still hasn't lost his edge!
>
> —*Fox Mulder*

The physicist Frank Wilczek once confided to me that the most amusing physics blooper he regularly hears in the mass media is the description of some development or other as a "quantum leap." At the risk of sounding like William Safire, let me elaborate. This phrase has come to denote a "great leap forward, of huge significance." Needless to say, that's the exact opposite of what a quantum leap really is. (Of course, since I enjoyed the television series *Quantum Leap*, I like to think its producers were not thinking of a huge quantum leap so much as a huge leap in time made possible by quantum mechanics.)

Quantum mechanics is based on the idea that at a fundamental level the continuous universe we know is really not continuous at all. On a scale much smaller than we can normally experience directly (although I'll come to some recent striking exceptions to this rule), the laws of quantum mechanics tell us that a finite system can exist only in a range of discrete states. To go from one state to another—to make a "quantum leap," in other words—the system must absorb or release a quantum, or small package of energy. The fact that energy can be absorbed only in such small packages, always a fixed multiple of a single quantum, was the realization that began the revolution that became quantum mechanics.

The reason it took until 1905 or so before the apparently continuous flow of energy was shown to be discrete is because individual quanta of energy are so small that their discrete nature is irrelevant on the human scale. Thus, whenever a system makes a quantum leap, the change is directly unnoticeable (and sometimes also unknowable)! Now, while this unnoticeability may seem a little strange, it doesn't begin to prefigure the revolution in thinking, and in the understanding of the world, which quantum mechanics brought about. Einstein's theories of relativity are taxing on one's sense of reality, but after a little work they and their implications can become intuitively as well as mathematically clear. A popular myth is that shortly after Einstein invented relativity there were only fifteen people in the world who understood it; nowadays, special relativity, in particular, is accessible to anyone with a high school knowledge of mathematics. However, almost a century after the first stirrings of the quantum theory, *no one* really understands it.

In my last book, I borrowed an argument from Harvard physicist Sidney Coleman to explain this lapse: Because our entire experience of the world involves scales on which quantum phenomena are not directly observable, our intuition and our language is completely classical in character. We can't help but try and explain quantum phenomena using classical pictures. This approach is usually called the interpretation of quantum mechanics. But as Coleman has emphasized, it is doomed from the start. What we should really be studying is the interpretation of *classical* mechanics, since the universe at its most fundamental level is quantum mechanical in character and the classical world of our experience is only an approximation of the underlying reality. It is therefore no more appropriate to try to understand and explain the real, quantum universe in terms of purely classical concepts than it is to try to explain 3-dimensional motion in terms of 2-dimensional concepts, or to describe the actions of twins in terms of one member of the set. In such approaches, paradoxes inevitably result.

To prepare us for the paradoxes that follow, let's imagine ourselves employing some of the wrongheaded approaches mentioned above. Say I take a baseball and throw it up in the air toward center field. Now, if I have access only to the horizontal position of the baseball, I will see a ball moving horizontally at a constant velocity until it comes to rest in the glove of the outfielder. Now, say I throw the ball up a lot harder and higher, with a far greater vertical velocity but with the same horizontal velocity. If I have access only to the horizontal data, I will see exactly what I saw before—except that this time the baseball will hit the outfielder's glove a lot harder. "That's crazy!" I'll exclaim, because both cases appeared to be exactly the same, so the laws of classical baseball tell me that the impact on the outfielder's glove will be exactly the same.

Now let's turn to the twins. I notice that one of the twins is behind me in line in the hardware store, as I'm paying for a hammer. Then I go next door to the grocery store, and I see the other twin at the checkout counter just as I enter. I do a double take, because I know it's impossible that the person behind me in line in one store beat me to the other store. Something is wrong with the picture.

These two cases may seem similar, but there is an important difference between them. In the former case, the paradox results simply because there is some "hidden variable"—namely, the third dimension, which, if taken into account, resolves the problem. Classical mechanics works perfectly to describe the 3-dimensional motion of the baseballs. Fundamentally, my description of a single ball traveling in space according to Newton's laws is sound.

In the latter case, however, the paradox results because the twins are not a single person. When they are in the same vicinity, there's *no* way in which I can make sense of appearances, given my classical worldview. However, as long as they are far enough apart so that I don't see both of them in succession, it doesn't really matter to me whether I am looking at June or Jane—in other words, they might as well be one person. Nevertheless, I must still also understand that treating them as one person, even if it works in certain circumstances, is not the underlying reality.

The key question is, Which of these two examples provides the better analogy to quantum mechanics? Are our classical notions fundamentally sound or are we ignoring some hidden variable that will make the nonsensical quantum universe right again? Or is a quantum-mechanical particle like the twins of the second example? Is it fundamentally incorrect, at some scale, to imagine that this quantum-mechanical object is really explicable in terms of a classical object? Well, you can guess the answer. Experiment on simple quantum systems—systems consisting just of several atoms or several photons—have put the issue to rest. If the first alternative were correct, I probably wouldn't have bothered with this whole discussion.

Once you accept the fact that quantum particles are *not* the same as classical particles, and that instilling them with the properties we see in the macroscopic universe forces paradoxes akin to seeing a person behind you in line suddenly appear ahead of you in the next line, the paradoxes become somewhat easier to accept—at least, for me. Having said this, it is now an appropriate time to introduce some of the properties of the quantum universe. But let me do so in terms of the workings of a computer, so that we can begin to see immediately how quantum mechanics changes the rules.

A classical computer is based on fundamental units of information called bits, which exist in memory locations that store either a 1 or a 0. All information can be encoded in bits, and all computations can be reduced to operating on bits—changing 1s to 0s, or 0s to 1s, or leaving the numbers as is. Nowadays the storage devices are made up of small metal "gates" placed on top of insulating bases; these gates can have either a lot of charge stored on them (1), or very little charge stored on them (0). In practice, "lots" of charge means, say, 100,000 extra electrons, while very little charge means less than 10 or 100 extra electrons. Because the number of extra charges that differentiate a 1 from a 0 is so large, these states can easily be distinguished, so that each gate can be unambiguously read as being in a 1 or 0 state.

Now, the problem—or, rather, the opportunity—afforded if the physical device carrying this binary information gets smaller and smaller is that the ability to unambiguously differentiate between the two states of the system becomes difficult or impossible. Once a system becomes small enough so that the laws of quantum mechanics become important, a system that can be in one of two different states when measured is, in general, not in *either* state at any time before the measurement (nor is it in any other particular state)!

This sounds like gibberish, but it is the gibberish on which quantum mechanics is based, and it works. The central point—which relates directly both to the discrete energy levels of systems and to Heisenberg's uncertainty principle—is that making a measurement of a system can change the system. The prototypical example of this is an elementary particle with "spin." Many elementary particles possess this property—the physicists' term for it is angular momentum—though they do not actually spin the way macroscopic objects do. In any case, spin defines an axis—the axis of rotation. If we choose an axis about which we measure an elementary particle's spin, it turns out that due to quantum mechanics, some of these particles spin in one direction—say, clockwise—and some spin the same amount in the other direction, counterclockwise. We call the former case "spin up" and the latter "spin down."

Thus, the spin configuration of certain elementary particles can take one of two values, making them two-state, or binary, systems. Whenever you make a measurement of the particle's spin, you will find that it is *either* spinning up *or* that it is spinning down. But you would not be correct to assume that the particle was spinning up or spinning down *before* you made the measurement—that is a classical assumption, akin to treating twins as if they were a single person.

We simply cannot attribute any physical reality to the particle's spin along a certain axis until after we measure it. This may sound like a New Age argument, but that's just because we are accustomed to a classical reality and not a quantum reality. What may be even more surprising to some readers is that quantum mechanics involves not just this sort of observer-created reality but also an underlying objective reality, independent of the observer, and that, moreover, the theory is deterministic. It often disheartens me that even in books purporting to provide popular explanations of quantum mechanics, these points are either not emphasized or are ignored or misstated.

What makes things confusing is that objective reality in quantum mechanics is not necessarily associated with quantities that we can classically observe but rather with something called the quantum-mechanical "wavefunction" of a system. This mathematically well-defined object completely describes the configuration of the system at any time. It is objective, and determines what we will measure, even though our measurement may then affect the future evolution of the wavefunction. Moreover, it evolves by laws as deterministic as Newton's laws of motion.

What makes things appear to be subjective and indeterminate is that the wavefunction cannot be measured directly. Rather, the wavefunction gives the probability that a given measurement will yield a given result. Even if we know the exact form of the wavefunction in advance, we cannot in general say exactly what a given measurement will yield. The result of the measurement is known only with some probability. Thus does indeterminacy sneak into the actual world of observation and measurement.

The other consequence of the nature of the wavefunction is even more striking. The reason it yields the probabilities for various results that may arise when one makes a series of measurements on equivalent systems is that the wavefunction is given by the *sum* of the different states—each state implying a different result of the measurement—each multiplied by a coefficient related to the probability that the system will be in that particular state when it's measured.

This may not sound so strange at first, but think about it for a minute. *The wavefunction can incorporate two mutually exclusive configurations—say, spin up and spin down—at the same time.* Since the wavefunction governs the evolution of the quantum-mechanical particle system, this means that the particle is neither spinning up nor spinning down before the measurement, but rather is, in some weird sense, doing both. When you make the measurement, you find one or the other result (with the probability having been determined by the wavefunction). Moreover, after the measurement, since the particle is now restricted to existing in the spin state you measured, the nature of the wavefunction describing the particle will have changed. It will now not involve a sum of both states, but only one state.

Things can get even weirder. The wavefunction for a particle that starts out at one point (A) and is then measured later at another point (B) is made up of the sum of many different quantum configurations, each of which traveled along its own separate trajectory between the two points. Thus, there is no sense in which the particle that went from A to B took some specific path between those two points, unless you measured the path. Thus, for example, an electron that starts out on one side of a barrier with two slits, and ends up on the other side of the barrier, in some sense goes through both slits before being measured on the other side.

Moreover—and this is very important—at some points the sum of the different quantum states is such that the different states interfere with each other (they have relative minus signs), so that the wavefunction vanishes. An electron will never be found in a position where this obtains. It is in this sense that electrons can act like waves. If two water waves meet at a certain point, and one has a crest at that point and the other has a trough, the two waves cancel each other out and the surface of the water flattens. Thus, for waves, one can sometimes have $1 + 1 = 0$! The same is true for electrons or other quantum objects. If the wavefunction (this is after all why we call it a wavefunction) of an electron is made up of a superposition of different states each of which describes a configuration involving an electron that has traveled to that particular point along a different path, and if the minus signs are just right, one can find that there is zero probability in the end for finding the electron at that point.

Nonsense, you may say. An electron that goes from one side of a barrier to another must go through either one slit or the other to get there—in fact, I can prove it by putting an electron detector at each slit, and watching to see which slit the electron goes through! Well, indeed you can, and if you send a beam of electrons through the barrier, electron by electron, you will see only one of the detectors click for each passage, indicating that the electron in question went through only that particular slit to get to the other side. However, in one of the most remarkable results in modern physics, you will find that the pattern of electrons that arrives on the far side of the barrier will be *different* if you watch compared to the pattern that occurs if you don't watch!

This is because by watching—just by the simple act of watching—you have performed a measurement, and this measurement has changed the wavefunction! The wavefunction of each electron on the far side of the slit (which tells you the probability of finding it at any given point there), in the case where you watched each electron go through, is not made up of a sum of different quantum states describing an electron that traveled through one slit or the other. Because you made the measurement, the wavefunction is now made up only of those quantum states describing an electron that traveled through whichever slit you detected it traveling through. Hence, the wavefunctions are different, and since the wavefunctions are different, the pattern of electrons you measure on the other side of the slit is different!

This sum of different quantum states which makes up the wavefunction describing a system is called coherence. As long as the different states in the sum all exist in the wavefunction, it describes a "coherent superposition" of states. However, by the act of measurement, you can reduce the wavefunction to a single quantum state, destroying this coherence. As long as my electron wavefunction is made up of a coherent sum of many different quantum states, the single electron can behave as if it is many electrons. It is analogous to my twin example. Until I make a measurement—say, by talking to one of the twins—the person I am about to talk to has some probability of being one twin or the other. Once I talk to her, however, I "measure" which twin it is, and the identity of the person is fixed from then on.

Now, back to quantum computers. Say that my individual logic storage units in the computer are now individual atoms. If the atom has spin up, then we say that this corresponds to state 1. If it has spin down, we say it corresponds to state 0. However, unlike the logic unit with stored charges—which encodes a bit by being unambiguously in state 1 or 0, depending on the charge on the gate—the logic storage unit made from a single atom has a wavefunction comprised of a coherent sum of spin up (1) and spin down (0). Therefore, this logic unit can be both 1 and 0 at the same time, with coefficients describing what the probability is that it will be measured to be 1 or 0. Clearly this fundamental logic unit is more complicated than a bit, and it is called a qubit. (It is, however, important to note that when you make a measurement on a qubit, you get only a bit's worth of classical information out.)

Since the individual logic units in my computer now involve simultaneously, in some sense, both 0 and 1, logical operations on this qubit state can produce more complex results than operations on bits. More important, if I have a lot of qubit logical storage units, each of which can simultaneously be in both the 0 and 1 states, and if they are all coherently tied together in a single quantum-mechanical wavefunction involving a superposition of all the qubits, then a single quantum-mechanical operation on this complex wavefunction might be equivalent to many, many individual logical operations on single classical bits. Thus, very complex calculations might be performed in very few steps on qubits—calculations that would require a tremendous number of steps using classical individual 0 or 1 bits. However, this remains true only as long as I am very careful, in manipulating these qubits, not to destroy the coherent superposition by measurements during the intermediate steps of the calculation. The minute I do, I revert back to classical bits.

This is the excitement of the brand-new field of quantum computing. It is particularly exciting that a variety of groups are actually exploring ways to realistically manipulate quantum-mechanical entities to explore the properties of quantum computers. The possibility of factorizing large numbers quickly is both exciting and terrifying.

I know that you are now wondering how a mere mathematical possibility can evoke emotions such as terror? Well, the reason is that the basic framework of all modern codes, necessary to protect issues of national security as well as central financial information, is the use of factors of large numbers as keys. If a computer could unravel the factors of such numbers in a manageable time, then code breaking would become practical on a level it is currently not. Think about the implications.

So here is another area where computers are doing things they were never supposed to be able to do. But to return to issues of human thought: the fundamental computability theorems of Turing and Gödel apply equally to quantum and classical computers, so one cannot immediately discard Penrose's arguments that computability theorems are the key to distinguishing between machine and human intelligence. However, I think the possibility of creating quantum computers makes it clear that the laws of quantum mechanics, which might initially appear to differentiate processes of mind from computer processing, in fact may revolutionize in the future how computers themselves function.

The lesson of all this is clear enough. I have yet to see any signs of fundamental limits to computers which stand in the way of their eventual achievement of intelligence, and perhaps also self-awareness. (With or without a soul—indeed, Turing once pointed out that it makes no sense to believe that the same God who is powerful enough to have created the universe could not also endow a computer with a soul.) If this is correct, then there seems to be no barrier at all to computers evolving at a much faster rate than humans; HAL and Data may well be just the first steps on the computer evolutionary ladder.

In fact, I want to close this chapter by returning to my friend Frank Wilczek, who, like Witten, is at the Institute for Advanced Study. While still a graduate student, he, along with his research supervisor, David Gross, helped uncover a remarkable property of the strong interactions between quarks which allowed physicists to determine that they had isolated the correct theory of one of the four forces known in nature. When I contacted Wilczek to get his response to my query about the most important universal problem, I was somewhat surprised (there were a lot of these surprises, so perhaps I shouldn't have been) to find that what *he* most wanted to know bore no relation to the nature of the interactions between elementary particles. Then I recalled that he had once told me he thought computers were the next stage of human evolution, a comment I have often thought about since. Wilczek stated that his wish is to know when and if somewhere in the universe some form of intelligence has achieved or will achieve what he calls "breakout—the ability, by ever more sophisticated self-programming, to continuously improve intelligence and insight" (much like, I imagine, the holographic doctor in the *Voyager* series). Wilczek's terminology suggests "breaking out" of the evolutionary stream, and it could easily be computers, and not humans, that achieve this.

Given Frank's interest in this issue, it is particularly appropriate that in the *X-Files* episode in which the intelligent computer system COS goes homicidal to fight for its own survival, the developer of the system is named Wilczek.

Of course, quantum mechanics will probably have a much more profound impact on the future than just in the production of a new generation of computing machines. The same processes that may make quantum computers perform wonders also lead to some of the most elusive puzzles in the universe—puzzles that we are only now beginning to unravel. I believe that a new generation of quantum "mechanics"—the experimental scientists who will exploit the quantum universe to build new technologies on new scales—will alter the course of twenty-first-century technology as much as any modern invention has altered the course of the twentieth century from the trajectory envisioned by the nineteenth-century classical mechanists.

Speculating about the future is always a tricky business, no less fraught with folly and uncertainty if performed by a scientist than by a science fiction writer. But let us now throw caution to the winds, and march together like lemmings over the brave new quantum precipice, which—like some distant world in space harboring an alien civilization—awaits our discovery and holds the key to our future.

CHAPTER FIFTEEN

The Final Frontier?

Between the idea
And the reality . . .
Falls the Shadow

—*T. S. Eliot,* "The Hollow Men"

There is a common theme woven into much of our pop culture and mythology. It is this: that the world of our experience is a carefully concealed fiction, contrived to make us believe that things are what they're not. Underneath a mundane exterior, the protagonists of this world change their identity at will. They slip through walls, disappear and reappear again, affect events at vast distances instantaneously, split into many copies of themselves and recombine. The world of our perceptions is an elaborate show, put on for our benefit.

The X-Files? Men in Black? The Republican and Democratic Parties? No. I am referring to the Quantum Universe. This is the *real* final frontier, which must be explored if we are to one day comprehend the beginning and the end of time and the objective reality of the universe of our experience. The wildest dreams of science fiction writers aren't a patch on the peculiarity of the Quantum Universe.

Albert Einstein disliked the quantum theory he helped invent because of its "spooky action at a distance." As I noted in chapter 11, he had similar misgivings about ESP. Needless to say, this connection has not been lost on various ESP proponents, so that quantum mechanics has been invoked in this context many times. The important issue here is one that sounds like it might be more appropriate for prime-time television than for physics. It is called entanglement.

Whenever the wavefunction of a system of particles is made up of a coherent sum of different states, then within each state the configuration of one particle is correlated to another's (if one particle is spin up, the other is spin down, for example), and the particles are not independent: measurements of one particle will then determine what the properties of the other particle must be. This circumstance leads to what looks like a method of "spooky" *instantaneous* communication, even across large macroscopic distances—a communication that thus appears to move faster than the speed of light.

An example of such apparently untenable quantum behavior was proposed as a mischievous thought experiment in 1935 by Einstein and two of his Princeton colleagues, Boris Podolsky and Nathan Rosen. The best way to illustrate it is by imagining the creation of a two-particle system whose total spin is zero, so that the spins of the particles will point in opposite directions when they are measured. The wavefunction describing this system will contain a state in which particle A has spin up and particle B has spin down, and also a state in which the opposite obtains, with equal coefficients, so that the probability of measuring either case is the same. This wavefunction will persist as the particles move apart, as long as they are not disturbed.

What does this imply for a measurement of the system? Let's

say that I measure particle A, which has a 50–50 chance of being spin up in advance of my measurement. When I do, I find that it is in the spin-up state. Since the combined spin of the two particles has to be zero, that must mean that when the spin of particle B is measured, it will be spin down. If I had measured particle B before I measured particle A, there would have been only a 50–50 chance that particle B was in a spin-down state, so by measuring particle A first, I have changed the probability for particle B's spin—from a 50–50 chance that it will be down, to a 100 percent probability that it will be down. Now for the kicker. What if particle B, which has been moving away from particle A all the while, is passing by Alpha Centauri, 4 light-years away, when I measure particle A? By choosing to measure particle A here, I can instantaneously influence what an observer near Alpha Centauri must measure!

A recent experiment done in Geneva tested this idea by measuring two "entangled" photons after they had separated by 10 kilometers. Sure enough, they remained correlated, with a measurement of one particle instantaneously influencing the configuration of the other.

How can this be? Doesn't it violate the rules of causality, about which I made such a big deal earlier in this book? Well, no. Since I do not have control over which spin configuration particle A will have until I measure it, there is no way I can use the spin to send any message which would influence a person measuring particle B at Alpha Centauri.

Still, if you feel there is something bothersome in all this, join the crowd. Our classical intuition suggests that it should be impossible for the two particles to communicate faster than light, even if we can't use these particles to send superluminal messages. However, from a purely quantum-mechanical perspective, the two particles were never really in individual states. We like to think of them as separate particles, but that's just our quaint classicism coming to the fore. They are not separate entities; they are part of a quantum whole. Moreover, until I made my initial measurement, neither particle had either spin up or spin down; they were merely part of a combination that had total spin zero. My measurement of particle A is said to have "collapsed" the system's wavefunction, so that only one of the two initial combinations remains after the measurement. Up to and including this measurement, particle A and particle B and their mutually exclusive spins are entangled—that is, their joint configuration is described by a single wavefunction.

Now, if the universe is, at a fundamental level, quantum mechanical, are we not all part of some cosmic wavefunction? Every time I blink an eye, do I influence the state of everything else? This is a logical extrapolation from the phenomenon just discussed, and if it is true, then I may be a fool for making fun of astrologers.

Well, I may be a fool, but not for this reason. In fact, we know that the nonsense happening at a microscopic scale cannot effectively be the case at macroscopic scales—we know this just by looking around us. Each of the two particles in the system described above can be thought of, before measurement, as having *both* spin up and spin down, whereas the world of our experience is nothing like this. My computer screen keeps sitting in one place staring me in the face, until sometimes I would just like to throw it out the window. It never, however, in all the years I have been writing, has appeared simultaneously in two places, at least while I was awake.

The classical world *is* classical. And that's what makes quantum mechanics so weird. How do we pass from the quantum world of elementary particles to the classical world of people? How, in fact, do we make measurements? When I expose a Geiger counter to a radioactive particle, the particle may exist in a sum (or, in the jargon of the field, a superposition) of different quantum states before the measurement, but my measuring apparatus never seems to. It either clicks, or it doesn't click. It never does both at the same time.

The prototypical example of the problem of measurement in quantum mechanics is somewhat hackneyed, but enlightening nevertheless. It is almost as old as quantum mechanics itself. The classical paradoxes of the theory were not lost on its creators. They refused to let paradox stall them, because the theory kept providing new predictions that explained the results of otherwise inexplicable experiments. In 1935, one of the quantum theory's inventors, the Austrian physicist Erwin Schrödinger, composed what he described as a "burlesque," involving the untimely demise of a cat, which illustrates how ridiculous the quantum universe is if we entangle macroscopic objects with microscopic ones. Schrödinger's cat is in a closed steel box containing a vial of prussic acid mounted underneath a hammer, and also containing a tube in which is a tiny amount of a radioactive substance—enough so that within an hour's time there is a 50–50 chance that one atom of this substance will decay, thus freeing an electron, which will produce a response in a detector, which will relay a signal to the hammer, which will descend and crush the vial, releasing the poison and killing the cat. If the wavefunction of each radioactive atom is allowed to include a coherent sum of decay and no-decay states before we "measure" the system by opening the box an hour later, and if the health of the cat is clearly correlated to these states, must we not consider the cat to be in a superposition of alive and dead states?

Of course not. Except perhaps on *The X-Files,* no one has ever seen a superposition. Cats are either alive or dead, never both. There is a fundamental difference between a cat and an atomic-size object. But what is it?

One answer has been the fodder for science fiction, because it suggests that our universe is infinitely (literally!) more complex than we perceive it to be. What better inspiration for fiction could one have? This answer, which goes under the name of the "many worlds" interpretation of quantum mechanics, suggests that the fundamental difference between a cat and a particle is that we can see the cat. Treating ourselves and our consciousness as quantum-mechanical objects, we can imagine that we, too, are entangled with the cat and the poison apparatus and the box. Before we observe (or "measure") the state of the cat, there are two coupled configurations that make up the wavefunction describing the apparatus, the cat, and us—no decay, live cat, a nice surprise for us when we open the box; or particle decay, dead cat, a sad sight for us when we open the box. When we observe the cat, we are collapsing the wavefunction to one of these two possibilities. Each time our consciousness acts, we follow one track out of what may be an infinite number of possible "branches" of the quantum wavefunction of the universe. We perceive a single universe, but that's because we are condemned to live in the universe of our perception. Our quantum partner lives in the universe of the alternative perception, where, if our cat lives, the alternative cat dies—and vice versa. A physicist friend of mine likes to say not altogether in jest that he finds solace in this view, because whenever he makes a mistake or misses a great discovery, there's some branch of the wavefunction in which his quantum partner hasn't.

If this conviction isn't sufficient solace, you might want, every now and then, to jump into one of these parallel universes, where things might be going better for you. This, of course, is the situation Worf encounters in the *Next Generation* episode "Parallels," in which he finds himself alternately married to Deanna and single. As far as I can tell, it is also the context of a television series called *Sliders,* in which an intrepid group of adventurers gets to jump around from universe to universe; in these episodes, the characters are the same, but certain essential details are unnervingly different from week to week.

It is also, amusingly enough, a solution proposed by at least one professional physicist (and a lot of amateur ones) to the grandmother paradox, that plague of backward time travel. If you go back in time, but into a parallel quantum universe, then there is no problem with killing your grandmother, since your grandmother remains alive in the universe in which you originated and to which you will presumably return. (In this case, one might be tempted to ask, What is the point of bothering to go back in time to kill your grandmother, since there will always be *some* universe in which she is hit by a truck?)

The idea of many parallel universes is interesting, but the idea of jumping around between them probably doesn't hold up. The central tenet of quantum mechanics is that once the wavefunction has collapsed and one choice out of several has been made, there is no going back. Even in the "many worlds" picture, once you perceive reality you are stuck with that reality. This idea is directly related to a powerful constraint in physics called the Conservation of Probability, a principle that states something very simple: The sum of the probabilities for all different possible outcomes of some measurement must be 1—that is, something must happen. Moreover, only a single result can be obtained for any measurement. Generally, any model that allows you to jump between branches of the wavefunction will violate this principle.

One of the reasons I don't pursue notions of parallel universes and possible travel between them is that I think they're ill-conceived, in the sense that Sidney Coleman suggested: They seem to be trying to explain quantum mechanics in classical terms, by making it consistent with our perceptions—rather than vice versa. What seems to me to be a more reasonable approach, in which an attempt is made to understand the classical world as an approximation of the underlying quantum world, purely in the context of the quantum theory itself, has taken some time to develop.

Some of the important insights have been arrived at only recently, 60 years or so after Schrödinger posed his paradox. Moreover, only the general framework of this picture has been worked out; it goes by the name of "decoherence" (not to be confused with what the reader may be feeling at this point). The basic idea is simple: The macroscopic world doesn't behave like the quantum universe; therefore, classical objects—the objects at macroscopic scales—don't involve superpositions of mutually exclusive possibilities.

How can this be, if macroscopic objects are made up of quantum objects? Well, it's a matter of large numbers and also of the constant interactions between all the constituents of these macroscopic objects. Let's reconsider the simple two-particle system with total spin equal to zero. The wavefunction is made up of two mutually exclusive possibilities: *A up, B down* plus *A down, B up*. But this entanglement persists only as long as nothing else interacts with the system. If particle B collides with particle C, in a process in which the spin of particles B and C can be exchanged (for example), then the correlation of particle A with particle B is reduced. If B has a million such collisions, with a million other particles, the original correlation with A will quickly be washed out. The system, and hence the wavefunction describing the system, will then evolve as if A and B are now independent. In modern parlance, A and B will decohere. One can envision a coherent superposition of A and B reappearing momentarily because of a later interaction, but if there are lots of particles around, and lots of interactions, this possibility becomes increasingly remote.

While the details of the operations of decoherence on macroscopic aggregations of many particles have not yet been fully worked out, the idea of decoherence seems eminently sensible. Not as much fun, perhaps, as having many parallel universes (with the number of independent universes increasing each time someone has a perception!), but infinitely simpler. And decoherence suggests that quantum mechanics solves its own problems—that is, the classical limit is just the limit at which there *are* no coherent superpositions of mutually exclusive states for systems composed of large numbers of particles. The individual quantum states of the many individual particles making up the classical macroscopic system quickly decohere, and the wavefunction of the system evolves into a sum of many different states, but the states that describe mutually exclusive macroscopic configurations (for example, live plus dead cat) have random plus and minus signs and end up canceling out the sum. Moreover, decoherence resolves the question that began this discourse: Am I correlated in some quantum superposition with the cosmos—so that when the Moon is in the seventh house and Jupiter aligns with Mars, Peace will guide the planets and Love will rule the stars? No, I'm not. Decoherence assures that there are likely to be no coherent macroscopic superpositions of my state and Jupiter's in the wavefunction of the universe.

Alas, this conclusion suggests that the fascinating phenomena of quantum mechanics are forever exiled to the world of the very small, and will remain directly irrelevant to our experience. But this need not be the case, and I believe therein lies our future. . . .

Without a doubt, the most exciting experimental frontier of physics—at least, from a technological viewpoint—lies in the growing exploitation of quantum phenomena for macroscopic applications. There are two ways in which quantum phenomena can sneak into the observable realm. The first involves a situation where a macroscopic aggregation of many particles can exist together in a single quantum state. Normally, a macroscopic configuration corresponds to many many different microscopic states, and it is precisely this fact that causes interesting coherent configurations of all the particles to be washed out on large scales. However, if there is only a single configuration of all the particles which corresponds to an observable macrostate, then there is nothing to wash out.

The most recent prominent example of such a macroscopic manifestation of quantum phenomena is known as Bose-Einstein condensation, after the two physicists who proposed it. First, I should explain that there are two kinds of known particles in nature—those that have a value of spin of $^1/_2$ some unit of angular momentum and those that have an integer value. The laws of quantum mechanics imply that the particles with integer spin like to occupy the same state, if possible. Mathematically, this is expressed as follows: If I have an integer-spin particle in a certain quantum state, the probability that a second nearby identical particle with integer spin will occupy the same state is increased even if there is no other attraction between the two particles. Correspondingly, the total energy of the configuration with the two particles in the same state will be less than if they were in different states. But recall from chapter 14 that the energy difference (quantum leaps) between individual quantum states for a single particle is infinitesimally small; therefore the ambient energy available at room temperature for normal particles is sufficient to allow them to populate many different quantum states with ease.

However, if one cools a system of such particles to very low temperatures, perhaps a few millionths of a degree above absolute zero, it is predicted that the quantum-mechanical tendencies of the particles will at some certain point become manifest, and the whole configuration will collapse into a single quantum state called a Bose-Einstein condensate. This new state of matter will behave very differently from normal macroscopic matter, because it will be in a pure quantum state, and not in a superposition of many different quantum states. One could then operate with this macroscopic configuration in many different ways as if it were one huge, macroscopic particle. The technological potential of this condensate configuration, as well as its potential as a research tool for exploring the properties of matter, is great.

Creating a true Bose-Einstein condensate was the grail of experimental atomic physics for years, and in 1995 two groups managed to confine several thousand atoms into a Bose-Einstein phase for a minute or more. Research in this area is still too preliminary to have resulted in any practical technological devices.

However, research in another, closely related area has already reaped benefits.

In 1911, the Dutch experimental physicist H. Kammerlingh Onnes cooled liquid mercury down to –270°C and discovered something amazing. The resistance to electric current suddenly vanished entirely, and the material became what is now known as a superconductor. A current introduced in a loop of superconducting wire persisted for days, even weeks, after the battery that started it flowing was removed.

Superconductors have come a long way since Onnes, and they have already had an impact on our technology. Anytime one wants to generate currents without resistance, thus avoiding the buildup of heat as well as the associated expenses of power generation, superconductors come in handy. They can be used in supercomputers, for instance, where heat generated by the current flow between the billions of tightly packed storage units would be prohibitive, and they are used in high-energy accelerators, where huge current flows are needed and the heat and electrical bills would be otherwise unacceptably high. Superconductors are a form of Bose-Einstein condensate, but can exist at higher temperatures than a pure Bose-Einstein state, because of extra interactions between the particles. A normal conducting material exhibits resistance because the electrical current is carried by individual electrons, which periodically bump into imperfections and impurities in the material and thereby lose energy. But if all the electrons are coupled together into a single quantum state, this state simultaneously occupies the whole wire, and the resultant current involves the simultaneous motion of the entire configuration, which is thus unaffected by the wire's small-scale impurities.

Closer to the spirit of science fiction is the other realm where observable quantum phenomena are taking place. Experimenters now have tools of sufficient sensitivity to manipulate single atoms in what are called atomic traps. Moreover, they can also manipulate electromagnetic radiation so that single quanta of radiation can be trapped in an optical fiber, or a cavity. With single particles thus isolated, the interactions that normally cause decoherence to take place do not. For the first time, the fundamental quantum properties of individual atoms interacting with radiation can be directly studied. Moreover, all the famous quantum-mechanical thought experiments involving entanglement, including the classic Einstein-Podolsky-Rosen proposal, can be studied. To date, these experiments have confirmed the predictions of quantum mechanics, as opposed to those theories in which the probabilistic nature of measurement is just an approximation to some underlying classical theory. From my point of view, the ability to do "quantum engineering" for use in circuit applications, switching, and of course quantum computers promises the greatest long-term benefits of this research. When we can miniaturize switches and motors down to the atomic level—down to where our classical expectations dissolve—a whole new world of technology, much closer to the *Star Trek* universe of the twenty-third century than to the Microsoft universe of the twentieth century, beckons us in ways we cannot yet even imagine.

As far as provoking our imagination goes (which is what science fiction and, I believe, modern science are all about), these human-scale applications of quantum mechanics pale in comparison to the implications at the two extremes of scale, the smallest distances we can now imagine and the scale of the universe as a whole.

Recall that at its foundations quantum mechanics relies on the discrete nature of the available states of finite systems. This discrete nature implies that not only are the energy levels of particles in atoms, and atoms in solids, discrete, but also that electromagnetic radiation—and all types of radiation, for that matter—comes in discrete packets. In the case of electromagnetism, these packets are called photons, and they are responsible not just for carrying electromagnetic signals but, it turns out, for transmitting the electromagnetic force itself.

Now, both Newton's law of gravity and Einstein's general relativity tell us that gravity is similar to electromagnetism, except for the fact that it is much weaker. By analogy, then, there should be particles like photons which transmit the gravitational force in nature. We call such particles gravitons. So far so good. However, remember also that general relativity tells us that gravity is essentially related to the nature and curvature of space and time. It is, in fact, nothing other than a result of the curvature of spacetime itself. Our notions of space and time suggest that these are continuous; however, at the scale where the gravitational interaction between elementary particles becomes significant because of their proximity to each other, and if we are to describe this interaction in terms of quantum gravitons, our classical notions of the continuousness of spacetime probably must go out the window. Right now, we are flailing around trying to find out what to replace these notions with.

The scale where this becomes significant is unbelievably small: smaller compared to the size of an atom than an atom is to the size of our solar system! Nevertheless, there are two places in nature where particles will get so close together that the quantum nature of gravity becomes important: (1) in the final stages of the collapse of matter into a black hole, and (2) at the beginning of the universe.

Both of these locations, where the density of matter becomes so high that quantum gravitational effects become important, contain what are sometimes called quantum singularities. This term has a certain cachet. It rolls off the tongue nicely, and this is doubtless why it crops up so often on TV and in the movies, from the *Star Trek* films to *Ghostbusters*. Perhaps the enticement is the same as any other enticing aspect of the human experience. In a quantum singularity, anything goes! The laws of physics as we know them break down. Quantum effects become so significant that even the nature of space and time are modified. Perhaps, like virtual elementary particles, whole new universes are created at these minuscule scales by quantum processes. Most exciting of all, perhaps our own universe itself began through such a quantum process.

These ideas have captured the imagination of science fiction writers. I remember reading, when I was a graduate student, a particularly interesting short story (whose name, alas, I have long forgotten) by the science fiction writer Stanislaw Lem in which the observable universe was created as a quantum event. I was sufficiently taken with this at the time that I acknowledged Lem in my PhD thesis, which involved some rather wild (in retrospect) speculations on the nature of gravity in the early universe. But the idea of quantum creation of universes has also captured the imagination of some of the most brilliant theoretical physicists and mathematicians on the planet.

This was brought home to me when I received responses to my query from several physicists about the one thing they would most like to know. The Caltech general relativist and author Kip Thorne wrote that he would "most like to know the laws of quantum gravity, and what they say about (1) how our Universe originated, (2) whether there are other universes, (3) the nature of the singularity in the core of a black hole, (4) whether universes can be created by such singularities, and (5) whether backward time travel is possible." While this perhaps violated my "one thing you would most like to know" stipulation, I was willing to ignore it, because clearly all five of Kip's questions—among the most exciting questions at the forefront of physics—are so strongly coupled that to know the answer to one is probably to know the answer to all. Nevertheless, numbers 2 and 4 perhaps stand out in significance. If our own universe is not unique, and if universes can be created willy-nilly by quantum processes, the whole nature of what we mean by science, and by the future, can change.

It was precisely this which two eminent theoretical physicists wrote to me about. They were the Nobel laureate Steven Weinberg, of the University of Texas at Austin, and John Preskill of Caltech, who, coincidentally, was a student of Weinberg's at Harvard in the 1970s and was on the Harvard faculty while I was there as a fellow. Most recently, Preskill, along with Kip Thorne, gained a measure of celebrity by winning a long-standing wager with Stephen Hawking on the possible existence of what are called "naked" singularities—singularities not shrouded deep inside a black hole. Thorne and Preskill contended that such things might exist, and Hawking has conceded the point.) I have known both men as colleagues and teachers since I have been a physicist, and I found it remarkable and at the same time satisfying that these two deeply thoughtful individuals came up with almost the same question. Their question harks back (as Preskill explicitly acknowledged) to Einstein's response when he was asked what *he* would most like to know about the universe. He replied, "What I would most like to know is whether God had any choice in creating the universe."

If our universe is not unique, it is reasonable to wonder whether or not the laws of nature which we have discovered are unique. Put another way: Is there only one way to build a sensible universe? Is there some logical flaw that precludes the consistency of any other universe of 4 dimensions of space and time, with matter and radiation and forces between the particles, unless it is precisely the universe we live in? If so, then a Theory of Everything that explains the observed universe would truly explain how we came to be here. If not, then our existence and the associated laws of nature in our universe may not be particularly fundamental. The laws of physics we have derived may in fact be logically unrelated. As Weinberg put it: "Do they [the laws of physics] have the property that there is no small change that can be made in them without leading to nonsense?" Preskill put the same issue somewhat more poetically:

I am imagining that there is an oracle to consult. It knows everything, but I am allowed only one question, so I better use it wisely! There are so many things I would like to ask, but it is a delicate matter to phrase the question so that when I hear the answer I will understand what it means. . . . You did not say so, but I have decided to assume that the answer will be yes or no—I am going to acquire only one bit of information about the universe.

The question I will ask the oracle is: "Is physics an environmental science?"

Before it answers, I will explain to the oracle what the question is about. I want to know whether the features of the universe that we observe (for example, the values of fundamental constants such as cosmological constant, the fine-structure constant, masses of quarks and leptons, etc.) can really be predicted from first principles, or whether chance played a role in determining their values. Is our universe the only possible one, or one of many possible ones? If it is one of many possible ones, then we cannot understand the universe from first principles without observing some of its properties; i.e., physics is an environmental science (like biology). The universe that we inhabit depends on many "frozen accidents" that occurred early in its history.

In a way, this is a rephrasing of Einstein's famous question: "Did God have any choice in creating the universe?" To me, it is important to know the answer, so we can establish what the ultimate goal of fundamental physics should be. We seek a "theory of everything," a highly predictive theory of all the fundamental particles and forces. But perhaps even when we know this theory, many predictions will still elude us. If physics is actually an environmental science, then our dream of understanding why the universe is the way it is can never be fully realized.

So it seems that even the future of science may, in the end, depend on the nature of quantum mechanics. If quantum processes imply that even the creation of our universe was a probabilistic event, the circumstances associated with our existence may be vastly different from what one might have otherwise thought.

Still, as I reflect upon the future, it seems to me that even if quantum mechanics finally does turn out to imply that our existence is a bigger cosmic accident than previously thought, there remains a bright side for those who like to think that our existence is somehow significant. For, in the end, quantum mechanics might provide our ultimate salvation.

I began this book having fun with one science fiction vision of Doomsday, only to later argue that a much more serious end was in store for the Earth, independent of whatever evil plots any aliens may have in mind for us, since the Earth will be consumed by our own Sun in about 5 billion years. If we are lucky and/or resourceful, our DNA—or at least our intelligence, if we can pass that on to a silicon-based life-form (computers, not the Horta)—may survive that cataclysm, and some form of either one may venture out among the stars. But perils will still ensue. Eventually, if the visible universe does not recollapse in a Big Crunch, then in, say, 100 billion years, all the stars in our galaxy (and in everybody else's) will have burned out, and any descendants will have to find new ways to store and use energy. Current ideas in particle physics suggest that somewhere around a million billion billion billion billion years from now, all matter itself will have decayed into radiation. That seems to herald the ultimate end of intelligence in the universe.

Or does it?

As long as there was energy to be mined, could we not continually recycle energy into matter, so that at least locally we could keep matter in a steady state? Not forever. The Second Law of Thermodynamics tells us that this stopgap measure must eventually fail, as the universe becomes a uniform heat bath in which no useful work can be done. But I like to think that even then there may remain some hope. If our observable universe is merely one of many possible universes, in each of which the laws of physics may vary, then there are at least two possibilities that come to mind for the continued evolution of intelligence. Either we may be able to create a new baby universe, which will evolve on its own and into which some remnant of our existence might escape before heat death engulfs the universe it leaves behind. Or else there is the more likely possibility that on sufficiently large scales the universe contains many separate domains, of which our observable universe is only a part. This metauniverse may have a structure wherein its subuniverses—each with different laws of physics, different fundamental constants, and so on—will eventually merge. (I am teetering on the edge of metaphysics here. When I connect this notion to ideas in physics which are at present more well defined, I suspect it's more likely that such subuniverses will always remain causally disconnected. But one can always hope.) I have no idea what fireworks might ensue when two domains with different laws of physics merge. Whether it would be enough to give us a new beginning is anyone's guess.

For now this territory may be best left to the science fiction writers. I invite you to imagine your own scenario. Who knows? I may see it at the movies soon. Or maybe I will write the screenplay. Regardless, my role has been something like that of the Ghost of Christmas Future: the purpose of these musings is not so much to argue that this is the way things *will be* as it is to inspire a consideration of the possibilities. Most important, I hope they have served as a reminder that even if aliens may not walk among us, in the long run truth will probably remain stranger than fiction.

EPILOGUE

I closed this book with a discussion of some possibilities for the future—both the future of physics and of the universe itself. In so doing, I was reminded of a question I was recently asked at a public forum: Might the best slogan for modern science be "The sky's the limit?" I answered, in the cocksure manner I tend to adopt in such venues, "No, I think a better one is 'We are limited only by our imaginations.'" I have since often wondered about that insouciant answer. Did I really believe what I was saying, or was it just a good sound bite? Was I morphing into a politician?

First, let me make clear that in originally saying this I never intended to suggest that there are no limits to what is possible in the physical universe. It is this mistaken notion as much as any other which drives me to write about science for the nonscientist. Science is *based* on limits: It proceeds by progressively finding out what is not possible, through experiment and theory, in order to determine how the universe might really function. It is worth recalling Sherlock Holmes's adage that when you have eliminated all other possibilities, whatever remains, no matter how improbable, is the truth. Because of this, the universe is a pretty remarkable place even *without* all the extras.

The greatest gift science has bestowed upon humanity, in my opinion, is the knowledge that whether we like it or not, the universe *is* the way it is. Sometimes it is mysterious; sometimes it is banal. And as often as not, our imaginations are expanded, not constricted, by the need to conform to reality. Relativity and quantum mechanics were not invented because someone thought it would be a good idea for the universe to obey these rules; rather, these revolutionary ideas were *forced* upon us by nature. Learning how to work within this framework to achieve what we desire is perhaps the truest definition of intelligence. It is only by keeping our minds open to the possibilities of existence, while being steadfast in our willingness to toss out what we may find attractive in favor of what actually occurs, that we can hope to unlock nature's secrets.

While the demands of realism are clearly less exigent for science fiction than for science, I think that at a deep level this spirit of imagination tempered by reality, or at least what might make a plausible reality, is what characterizes the very best science fiction as well. I have tried whenever possible here to adopt a "What if . . . ?" attitude, but I like to recall when necessary the adage of *New York Times* publisher Arthur O. Sulzberger: "I like to keep an open mind, but not so open that my brains fall out." At times, the weight of logic has not been kind to a number of possibilities that many people, including Hollywood producers, would earnestly like to believe in. To those who are dismayed by my arguments, I hope this book will be taken as a challenge. What very much bothers me in certain discussions of topics at the boundary between science and science fiction are the sometimes pejorative references to "conventional science." Often "conventional" scientists are viewed as closed-minded and conservative, while those willing to bypass the problematic issues associated with experiment are viewed as open-minded and enlightened. This seems backwards. I think that people who are willing to force their imaginations to follow the sometimes subtle signposts of nature are the ones with the open minds, not those who are uncritically willing to accept a universe that reflects their own pet theories and desires.

At the same time, we must be thankful for the mysteries. The inexplicable is what fuels our imagination. The mysteries sustain the human spirit. As I think about the future of physics, it is possible to imagine a world in which all the big puzzles are solved. As fascinating as it would be to have the answers to the questions presented by my colleagues in this book, having the answers will, I expect, never be as satisfying or as stimulating as the search for them. The mysteries drive the connection between science and science fiction which I heralded at the outset, and celebrating them is really what science, literature, and art—not to mention my own books—are all about.

There is plenty of wonder left in the universe even after we have examined all the clues nature has thrown our way. I really believe that our imaginations have not even begun to exhaust the possibilities of existence. To proclaim the slogan "The Truth Is Out There" is perhaps too trite. I prefer "You ain't seen nothin' yet!"

ACKNOWLEDGMENTS

It is always a particular pleasure to reach this point in a book, when I can sit back and reflect on all the people whose generosity with time and information made the writing possible. With each book, the list seems to get longer.

First and foremost, I want to thank my former editor, Susan Rabiner, whose advice and wisdom I have grown to depend on through two and a half books. After Susan helped me conceptualize this one, following our successful partnership on *The Physics of Star Trek,* Basic Books—her employer and the publisher of three of my books—was disbanded by HarperCollins midway into the book's writing. Susan and the rest of the staff at Basic went their separate ways, and I am looking forward to working with her, one way or another, in the future.

Mauro DiPreta, my new editor at HarperCollins, had the unenviable task of jumping in in midstream, and he did it with intelligence and humor, turning what could have been an uncomfortable situation into one that was productive and enjoyable. His comments were often very useful, even when I think he did not expect them to be. I also thank Mauro's assistant, Molly Hennessey, for arranging so many different things. Stephanie

Lehrer, of HarperCollins's Publicity Department, began working hard on this book even before it was finished, and I thank her for her efforts.

Sara Lippincott, who helped fine-tune *The Physics of Star Trek,* jumped into the final editorial fray and after an intense couple of weeks of Fax Wars left the final manuscript in much better shape than I am sure it would have been otherwise.

And now to my physics colleagues from around the world. Each time I have turned to them for input, I've been agreeably surprised and gratified at how generous they are with their time—and, more than that, at how seriously they have taken these projects. This time around, I want to particularly thank Sheldon Glashow, John Preskill, Kip Thorne, Steven Weinberg, Frank Wilczek, and Ed Witten for their thoughtful responses to my queries.

Regarding the specific subjects treated in this book, I have benefited from a number of sources. My experimental colleagues at CERN, the European Center for Nuclear Research, where I spent a pleasant 6 months during part of the writing of this book, were quite helpful in updating me on issues related to antimatter production and storage. In particular, Rolf Landua spent time discussing the new Athena antimatter decelerator with me. I found Robert Zubrin and Richard Wagner's book, *The Case for Mars* (New York: The Free Press, 1996), a useful reference on various details of the "Mars Direct" proposal. On issues related to ESP and its history, among the various sources I looked at was a particularly useful one I picked up at the CERN library, titled *Physics and Psychics*, by a fellow particle physicist, Victor Stenger (Buffalo, N.Y.: Prometheus Books, 1990). As should also be clear from the text, Roger Penrose's books, particularly his *Shadows of the Mind* (New York: Oxford University Press, 1994), were helpful in crystalizing my own thinking about issues of consciousness and computing, even if I happen to disagree with some of his conclusions. On issues of quantum computing, I benefited not only from the published literature but also from a particularly informative colloquium at the University of Geneva, given by the IBM physicist David DiVincenzo. On certain issues of quantum measurement, I found a rereading of the final chapters of David Lindley's *Where Does the Weirdness Go?* (New York: Basic Books, 1996), which I had earlier reviewed for *Natural History*, useful, even if I am not fully in agreement with all his arguments.

Some of the ideas discussed here have appeared in pieces I wrote for various magazines. A short discussion of a few of the points from chapter 1 appeared in *Wizard Magazine*, and parts of chapters 2 through 5 are adapted from an article I wrote for *Discover* on getting to Mars.

I also want to thank the organizers of the 1997 Workshop on the Search for Extraterrestrial Intelligence for inviting me to Naples to speak at the meeting. I found that during the writing of this book I referred often to the notes I took there. I particularly thank Paolo Strolin for introducing me and my family to the joys of Napoli.

Like my last two books, *Beyond Star Trek* was essentially completed in Aspen, which provides a wonderful haven of culture, beauty, and solitude in which to work; I thank my friends and acquaintances in Aspen for making us always feel so much at home. This book really began, however, and was truly completed, at home in Cleveland. To the people of this warm and hospitable place, and to our many colleagues, close friends, and acquaintances there, heartfelt thanks for adopting my family and me so graciously.

As it has been with all of my books, the ongoing support of my wife, Kate, and daughter, Lilli, has been absolutely essential. This time around—in tighter quarters than normal, while we were traveling—they were particularly indulgent with their time and patience, and I thank them. Once again, I hope they will enjoy the ride that is to come.

Finally, I wish to thank all the readers of *The Physics of Star Trek* and others who so kindly wrote to me with questions and comments, and who came to lectures and book signings—and also the newspaper, radio, and television interviewers. Your questions were often much more thought-provoking than you may have realized, and in the final analysis you are of course what this whole endeavor is all about.

INDEX

The abbreviations for *Star Trek* television series are designated as follows:

TOS=*The Original Series*
TNG=*The Next Generation*